Flavor of Dairy Products

ACS SYMPOSIUM SERIES **971**

Flavor of Dairy Products

Keith R. Cadwallader, Editor
University of Illinois

MaryAnne Drake, Editor
North Carolina State University

Robert J. McGorrin, Editor
Oregon State University

Sponsored by the
Division of Agricultural and Food Chemistry, Inc.

American Chemical Society, Washington, DC

The paper used in this publication meets the minimum requirements of American National Standard for Information Sciences—Permanence of Paper for Printed Library Materials, ANSI Z39.48–1984.

ISBN: 978–0–8412–3968–5

Distributed by Oxford University Press

PRINTED IN THE UNITED STATES OF AMERICA

Foreword

The ACS Symposium Series was first published in 1974 to provide a mechanism for publishing symposia quickly in book form. The purpose of the series is to publish timely, comprehensive books developed from ACS sponsored symposia based on current scientific research. Occasionally, books are developed from symposia sponsored by other organizations when the topic is of keen interest to the chemistry audience.

Before agreeing to publish a book, the proposed table of contents is reviewed for appropriate and comprehensive coverage and for interest to the audience. Some papers may be excluded to better focus the book; others may be added to provide comprehensiveness. When appropriate, overview or introductory chapters are added. Drafts of chapters are peer-reviewed prior to final acceptance or rejection, and manuscripts are prepared in camera-ready format.

As a rule, only original research papers and original review papers are included in the volumes. Verbatim reproductions of previously published papers are not accepted.

ACS Books Department

Contents

Biochemical Aspects

Processing and Storage Effects

Indexes

Preface

In the past, the stability of milk and milk products was the primary consideration in dairy research, but this is no longer the principal objective due to the evolution of modern sanitary practices and effective pasteurization processes. Today, the manufacture of dairy products with consistently good flavor and texture has greater emphasis. In previous flavor studies, researchers had identified hundreds of volatile com-pounds, with little or no attention paid to their sensory contribution to the overall flavor of dairy products. The availability of powerful chrom-atographic separation techniques such as high-resolution gas chrom-atography in combination with mass spectrometry and olfactory detection ports has revolutionized the work on characterization of dairy flavor. Additionally, recent developments in sensory methods and our increased knowledge about the genomics of dairy culture organisms have allowed great advancements in our understanding of dairy flavor chemistry.

This book was developed from the symposium *Chemistry and Flavor of Dairy Products*, which was held at the 228th American Chemical Society (ACS) National Meeting in Philadelphia, Pennsylvania, August 22–26, 2004. It reports the latest developments in the evolution of dairy flavor research and presents updated information in the areas of instrumental analysis, biochemistry, processing, and shelf-life issues related to the flavor of dairy products.

We are grateful to the authors for their contributions that made publication of this book possible. We especially thank the ACS Division of Agricultural and Food Chemistry, Inc. and Dairy Management Inc. for their financial support.

Keith R. Cadwallader
Department of Food Science and Human Nutrition
University of Illinois
1302 West Pennsylvania Avenue
Urbana, IL 61801

MaryAnne Drake
Department of Food Science
Southeast Dairy Foods Research Center
North Carolina State University
Raleigh, NC 27695

Robert J. McGorrin
Department of Food Science and Technology
Oregon State University
100 Wiegand Hall
Corvallis, OR 97331

Overview

Chapter 1

Sixty Years of Research Associated with Flavor in Dairy Foods

W. James (Jim) Harper

Department of Food Science and Technology, The Ohio State University, Columbus, OH 43210

Prior to the rediscovery of chromatography in 1944, research in flavor chemistry was limited to "wet chemistry", which severely limited the scope of such research. The 1940's and 1950's saw the introduction of diverse chromatographic methods, radioisotope tracers, and mass spectrometry that permitted research in dairy foods flavor not previously possible. I was privileged to "grow-up" professionally during this period, and have used all of these methods during the past sixty years to explore the flavor of butter, cheese, milk, ice cream and fermented products. This chapter recounts my main experiences from the earliest work with amino acids as important compounds in the flavor of Cheddar cheese in the 1940's to high flavor impact polyfunctional thiols as potential flavor components in Cheddar cheese in the 2000's. In between there was research on flavor of other cheeses, hydrolytic rancidity in milk, sun-light flavor of milk, cooked flavor of milk and flavor of buttermilk

Progress in flavor research is dependent upon available methods and new methods lead to new discoveries. Prior to the rediscovery of chromatography in 1944 research in the field of flavor chemistry was limited to single classes of compounds, which severely limited the scope of such research. The 1940's and 1950s saw the introduction of diverse chromatographic methods, radioisotope tracers and mass spectroscopy that permitted research in dairy foods flavor not previously possible. I began work as a Lab Technician at Purdue University in 1942, and was privileged to "grow-up" professionally during this period of new method development, and used many of these methods during the past 60 years to explore the flavor of butter, cheese, milk, dairy ingredients and fermented products.

Tables outlining the methods developed over the past sixty years and in my own research program during that time will serve to introduce each decade.

1940-1949

The methods that became available during this decade and those used in my own research program are shown in Table I.

Table I. Methods introduced and those used my own research program during the decade 1940-1949

The World		*My Research*	
Year	*Method*	*Year*	*Research*
1944	Rediscovery of chromatography	1942-1943	Rapid colorimetric method for insoluble fatty acids in butter
1943	Microbiological Assay of amino acids	1946 1946-1949	Unsuccessful attempt to use paper chromatography for quantitation of amino acids; Application of microbiological assay for free amino acids in cheese.

I began my research as a student and laboratory technician in the lab of Dr. Bill Epple at Purdue University in 1942, working on a colorimetric method for long chain fatty acids in butter as a rapid means of following rancidity in gathered sour cream. The only method available at that time took 6 hours.

Upon completion of my BS degree in 1946, I started in graduate school at the University of Wisconsin under the direction of Dr. Hugo Sommer and Dr.

Arthur Swanson. Because I knew the least about cheese of all the dairy products, that area was chosen for research. I have continued some cheese ripening research ever since. The first work at Wisconsin was not on flavor, but on determining the nature of the white specks that formed in the interior of aged Cheddar cheese (*1*). This has been of interest to researchers since the late 1880's. In 1946 workers at Cornell thought that the white specks were tyrosine, whereas workers at Illinois claimed that they were calcium lactate. Using microbiological assay for amino acids and X-ray diffraction for calcium lactate, it was shown that both tyrosine and calcium lactate could be present, and in one case both were present in the same "white speck". One cheese had only leucine. Thus any compound that exceeded its solubility product could form the "white specks" (*1*).

Work then turned to the determination of free amino acids in Cheddar cheese during ripening, using a newly developed technique of microbiological assay. The research showed that for the 14 amino acids, which could be measured, the total water extractable free amino acid concentrations increased with age as shown for selected amino acids in Table II (*2*).

Table II. Effect of age on selected free amino acids in Cheddar cheese (data from reference *2*).

Age (weeks)	*mg/10 g cheese*			
	Glutamic acid	*Leucine*	*Valine*	*Lysine*
4	56	20	11	20
11	100	46	22	45
21	136	64	36	56
31	173	69	57	69

It was shown also that there was a definite relationship to the intensity of cheese flavor (*2,3*). This was especially true for glutamic acid, valine, isoleucine and leucine. Addition of the free amino acids in concentrations found in the cheeses to a bland, washed cottage cheese base was judged as being "cheesy" by a panel of judges familiar with Cheddar cheese flavor. The degree of "cheesy" increased with increasing age of the cheese and increasing concentrations of free amino acids. Other observations made during this study were that di- and tri-peptides gave a significantly greater growth response than the amino acid alone and that the rate of free amino acid production was greater for raw than for pasteurized Cheddar cheese (*2*). For example, the average free isoleucine content of 6 raw and past milk cheeses with a flavor intensity of 2 out of 4, the raw milk cheese was 60 mg/10g of cheese and the pasteurized milk cheese had 40 mg/10g of cheese.

Upon completion of my PhD degree, I started at the Ohio State University in 1949, partly as an Assistant Professor and partly as an agent for the USDA.

1950-1959

The methods that became available during this decade and those used in my own research program are shown in Table III.

Table III. Methods introduced and those used my own research program during the decade 1950-1959

The World		*My Research*	
Year	*Method*	*Year*	*Research*
1950	Radioactive tracers become commercially available	1950	Developed a column chromatographic method for low molecular weight fatty acids in Italian cheese
1952	Gas chromatographic equipment available	Mid to late 1950's	Applied many different chromatographic methods Built first gas chromatograph
Mid to late 1950's	Mass spectrometry is in its infancy, with much promise for flavor research	1957-1959	The use of radioactive tracers first used for CIP studies were now applied to cheese ripening research

Initially, research at OSU centered around the inability of the Italian cheese industry to produce their normal flavor in Romano and Provolone cheese, which was traced to a FDA embargo on imported rennet paste. The objective of the project, funded by the USDA, was to find out why rennet paste was important to the flavor of these cheeses. Tasting cheese with and without rennet paste suggested that lower molecular weight fatty acids were involved. Research very quickly showed that the problem was associated with the loss of lower molecular weight fatty acids, when cheese was made with rennet extract of purified rennet paste (*4-6*). This necessitated the development of a chromatographic method that could provide quantitation of the lower molecular weight fatty acids. Paper chromatography was tried and discarded. A simple column method, very crude by today's standards, was developed (*7*). The column is shown in Figure 1.

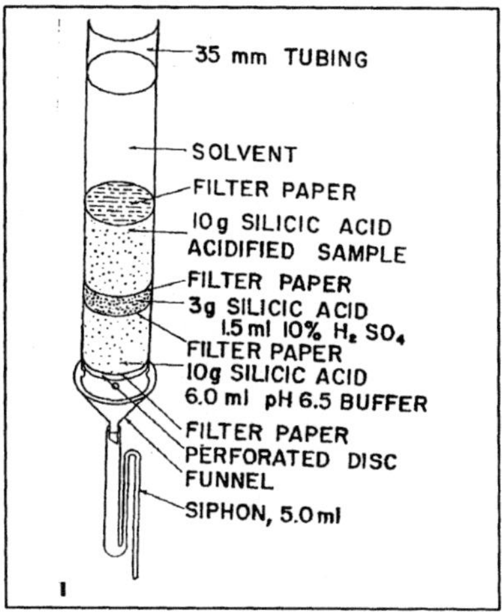

Figure 1. Diagram of a silica gel chromatographic column for determination of fatty acids in cheese. (Reproduced with permission from reference 7. Copyright 1953 American Dairy Science Association.)

We started out collecting 5 ml fractions by hand, and then built our own spring-loaded fraction collectors for $25 each, since we couldn't afford the commercial ones on the market. The method was applied to follow the ripening of Italian cheese showing the importance of butyric, caproic, caprylic and capric acids to the flavor of Romano and Provolone cheese (*8-10*). Also the method showed a selective liberation of lower molecular weight fatty acids from milk fat by a lipase in rennet paste that was also found in a gland at the bases of the tongue of ruminant animals (*11-13*). The mole percent of butyric acid liberated from milk fat was 13.5, 10.7, 32.8 and 44.4% for milk lipase, purified calf rennet paste, imported crude kid rennet paste and kid oral lipase, respectively (*12*). Partially, as a result of these findings the Italian cheese industry was able again to develop characteristic flavor and went from strength to strength.

During the early 1950's, the dairy industry was undergoing major changes that include the introduction of bulk tank pick up of milk, wide use of high temperature short time (HTST) pasteurization and homogenization together with a shift from home to store purchase of milk. Together these changes made lipolysis a major problem for the fluid milk industry. It was natural to modify the chromatographic method to provide a rapid, sensitive method (*14-16*) for studying these problems and to seek solutions for milk and milk products (*17-21*).

At the same time, other chromatographic methods were being developed or adapted and applied to cheese and other dairy products included ion exchange chromatography for free amino acids (*22-23*), paper chromatograph for carbonyl compounds (*24-29*) and our first attempt at a gas chromatographic (GC) method for fatty acids and amines in a home made GC that used an automatic titrimeter as detector, as shown in Figure 2. (*30-32*)

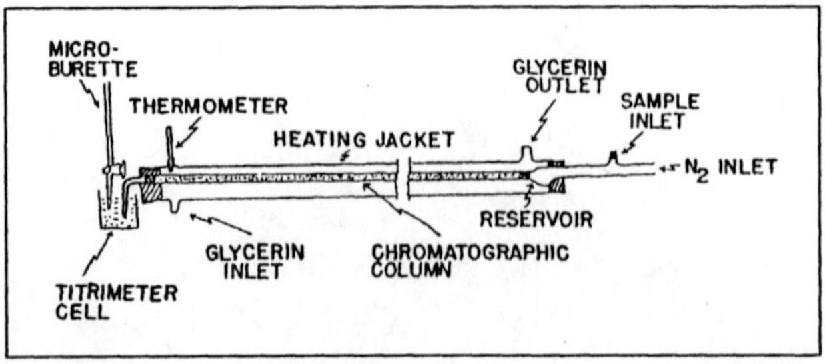

Figure 2. Diagram of Gas chromatographic set-up for measuring fatty acids and amines (Reproduced from reference 30. Copyright 1958 American Dairy Science Association.)

Work was instituted on the application of radioactive labeled compounds to begin to understand some of the mechanisms involved in cheese ripening, using a simple cheese slurry system to facilitate the use of the radioactive C^{14} labeled compounds (*24*). The slurry system used cheese curd at pH 5.3, moisture content of 60% and the same salt to moisture ratio as in the cheese under investigation.

For amino acid studies, in the beginning we made out our uniformly labeled C^{14} amino acids with bean sprouts and isolated the free amino acids by ion exchange chromatography. This was abandoned when commercial sources of amino acids became available.

1960-1969

The methods that became available during this decade and those used in my own research program are shown in Table IV.

This decade saw a continuation of studies on cheese and milk with an emphasis on the use of radioactive labeled compounds to explore avenues not otherwise available.

With homogenized milk becoming increasingly popular at this time, sunlight (light induced) flavor had become a major issue. The consensus of opinion was that the oxidation of methionine in casein to methional was the responsible compound for sunlight flavor. Adding S^{35} labeled methionine to milk, exposing it to light and trapping the different classes of volatile sulfur compounds

Table IV. Methods introduced and those used my own research program during the decade 1960-1969

The World		*My Research*	
Year	*Method*	*Year*	*Research*
1960-1969	GC systems become widely used – first publication related to cheese flavor in 1968	1960-1969	Increased use of radioactive materials for research (C^{14}, S^{32}, H^{3})
Late 1960's	HPLC introduced	1960-1969	Continued use of different chromatographic methods, beginning to used commercial GC
1960-1969	GC/MS coming into increasing use for flavor research	1969	Working with cheese slurries to accelerate ripening

released into a series of traps (Figure 3), it was possible to show that the methionine was degraded to mercaptans and sulfides without the formation of methional (*33*).

Further work was done with S^{35} cystine and with milk labeled through the injection of S^{35} cystine into a cow and collecting the milk. The same system as diagrammed in Figure 3 was used for this study. The relative amounts and ratios of the various classes of sulfur compounds were found for both the radiolabeled cystine and milk. Also exposure to direct sunlight for 30 minutes or to 1000 foot candles of fluorescent light for four hours gave essentially the same results as shown in Table V.

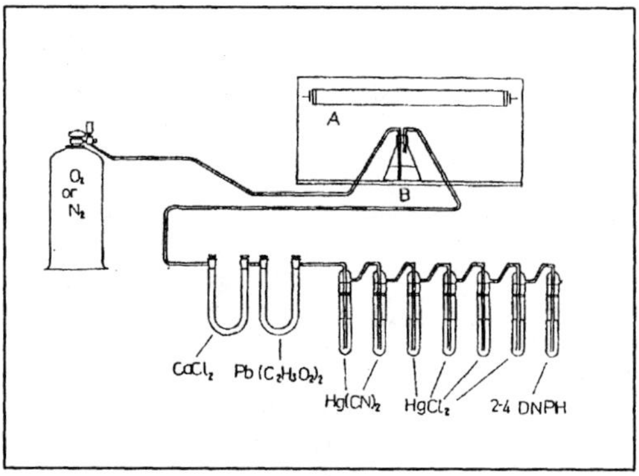

Figure 3. Diagram of system to collect volatile sulfur compounds during light treatment of fluid milk (Reproduced from reference 33. Copyright 1961 Agrar-Verlag Allgaeu GmbH, Kempten, Germany.)

Table V. Effect of fluorescent light and sunlight exposure on the formation of volatile sulfur compounds from S^{35} labeled milk. (Data from reference *44*)

Compound	*ppb S found based on specific activity of cystine S^{35}*	
	Fluorescent light	*Sunlight*
Hydrogen sulfide	0.7	0.9
Mercaptans	0.3	0.7
Sulfides	1.7	1.9
Carbonyl sulfur compounds	0.07	0.2

The distribution of the various classes of sulfur compounds were similar both in respect of exposure to sunlight or fluorescent light. The major class of compounds was sulfides, as would be expected, along with a small amount of carbonyl containing sulfur compounds. Paper chromatography showed the presence of two unidentified compounds. Methional was not detected in this study.

Another area of research centered on S^{35} labeled milk using sodium sulfate, sodium sulfide or barium sulfide introduced into the rumen of a cow through a canula or by the injection of S^{35} labeled methionine or cystine. The cow was maintained at the Ohio Agricultural Research Develop Center(OARDC in Wooster and radioactive milk obtained over a several year period of time. In addition to studying the time sequence for the labeling of individual milk proteins (*35-37*) the radioactive milks were used to study heat induced changes during the heat treatment of milk, cream and skim milk (*35-41*), as well as the utilization of sulfur amino acids by starter cultures and in cheese and fermented products (*37,42-44*)

The introduction of ultra high temperature (UHT) milk in Sweden in 1960 created a renewed interes in cooked flavor. Cooked flavor in milk was known at that time to involve the formation of sulfur compounds during the heating process, especially hydrogen sulfide. The radioactive milk using S^{35} barium sulfide showed a maximum specific activity for caseins and major whey protein at 24 hours and maximum specific activity for the milk fat globule membrane proteins at 36 hours (*33-35*). There was no labeling of the fat globule membrane proteins at 24 hours. Heating the 24 hour skim milk and cream to 90°C showed 1.5 times more S^{35} in the cream than the skim milk, but with a 50% lower specific activity in the cream than skim milk. Heating the 36 hours skim milk and cream showed a similar distribution of hydrogen sulfide in the two products, but with much higher specific activity in the cream than in the skim milk. This showed that both the whey protein and the fat globule membrane proteins contributed to the formation of hydrogen sulfide in the heated cream (*35, 37-38*).

As an outgrowth the study of carbonyl compounds in cheese and cultures, investigation was made of the ability of the fermentation system to utilize different substrates, including acetate, citrate, succinic, malate, glyoxylate and oxalacetate. The utilization of uniformly C^{14} labeled was followed in either broth cultures or in cheese slurries as outlined previously. The utilization of the various carbon substrates was essentially the same in thermophylic cultures in Romano, Provolone and Swiss type cheeses. Table VI shows selected results for utilization of the various substrates by *Streptococcus thermophilus* and in Romano and Swiss cheese slurries.

All four cultures investigated converted the various substrates into the same compounds. Citrate was converted to the following acids: isocitric, oxalsuccinic, alpha-ketoglutaric, alpha-acetolactic, glyoxylic, succinic and oxalacetic. Succinate was converted to alpha-acetolactic and oxalacetic acids. All thermophylic cultures also showed the formation of C^{14} glutamic acid, leucines and valine from the labeled citric acid.

Table VI. Utilization of C^{14} substrates by culture and cheese slurries

Substrate	*Percent Utilization*		
	S. thermophilus	*Swiss*	*Romano*
Citrate	61	67	59
Succinate	29	34	32
glyoxylate	11	10	10
Oxalacetate	2	1	1
Malate	1	1	1

1970-1979

The methods that became available during this decade and those used in my own research program are shown in Table VII.

Funds for dairy research became much more difficult to obtain, especially for flavor research. During this decade the only support for dairy flavor research came from OARDC Hatch funds. At the same time, funds for new equipment disappeared. At the beginning of this decade, we couldn't afford GC or HPLC equipment. However, funds for environmental research became readily available and much of my research shifted to this area.

Flavor related research focused on cheese ripening with an emphasis on accelerated ripening of cheese using the cheese slurry approach developed at The Ohio State University (*45*). These studies included protein degradation (*46*), free fatty acid development (*47-48*), carbohydrate metabolism (*49*), esterase activity (*50*) and amino acid catabolism (*51-54*).

Table VII. Methods introduced and those used my own research program during the decade 1970-1979

The World		*My Research*	
Year	*Method*	*Year*	*Research*
Early 1970's	Capillary GC introduced	Middle 1970's	Funds for dairy flavor research dry-up
1970-1979	Explosion of identification of volatile compounds in food products through application of a wide number of different techniques	1970-1979	Research started to turn away from measurement and identification of flavor volatile compounds, shifts to cheese slurries and mechanism involved in flavor changes
1970-1979	Methods for analysis and identification of flavor volatile are continuously improving	Late 1970's	Increasing "laundry lists" of volatile flavor compounds, without knowing which are most important becomes frustrating

1980-1989

The methods that became available during this decade and those used in my own research program are shown in Table VIII.

In 1981 I received an opportunity to focus entirely on dairy research at the New Zealand Dairy Research Institute and retired from The Ohio State University to accept this challenge. The research work was both on the physical functionality and flavor of dairy products as food ingredients. Most of this work was proprietary and published only in reports. This included flavor research on milk proteins and cheese, none of which was published. One study does merit mention, because of its significance to interpretation of research relating to cheese volatile compounds. Nine different methods of extracting volatiles were investigated with a three-year-old Cheddar cheese. These included vacuum distillation, supercritical extraction, and static and dynamic headspace methods, as well all the recognized solvent extraction systems. About 300 compounds were identified by GC/MS. However, only free fatty acids were common to all of the methods. A look at the results overall suggested that we were not looking at the same cheese. This means that extreme caution must be used in extrapolation from a given method of obtaining the volatiles from a dairy product.

Table VIII. Methods introduced and those used my own research program during the decade 1980-1989

The World		*My Research*	
Year	*Method*	*Year*	*Research*
Entire decade	Flavor chemist widely using GC/MS	1981	Enter first retirement and move to New Zealand
Mid to late 1980's	Olfactory GC used to provide a better estimation of the importance of different compounds to flavor	1981-1989	My flavor research work in New Zealand unpublished, except in reports
Late 1980's	Descriptive sensory methods beginning to get more attention	1987-1989	One study with 9 methods for obtaining volatiles emphasized that the method of extraction is critical

1990-1999

The methods that became available during this decade and those used in my own research program are shown in Table IX.

I received an invitation to return to the faculty of the Department of Food Science and Technology at The Ohio State University in 1992 and was appointed to the J. T. Parker Endowed Chair in Dairy Foods in 1993, where opportunity was given to "get back into" research related to the functionality of dairy products, including flavor. With respect to flavor, two main areas received attention: (a) Electronic nose research and (b) Swiss cheese research.

In 1994 the first commercial electronic noses (EN) were introduced into this country. My thoughts were that these were either going to be of significance to flavor research or were " the greatest scam that ever came down the road". To test this hypothesis, I obtained all of the instruments available at that time. Using one of the instruments that used organic polymer sensors, we differentiated cheeses, but it turned out that the basis of differentiation was on the moisture content of the cheese. Electronic noses worked with either organic polymers or metal oxide sensors extremely well in some applications and failed in other applications (*56-62*). The alcohol in beer or wine and the acetic acid in salad dressing used up so much of the signal that other flavor components could not be differentiated. This was overcome in large part when mass spectrometry began to be used as a sensor , using abundance of a given mass as a sensor element (*61-62*).

Table IX. Methods introduced and those used my own research program during the decade 1990-1999

The World		*My Research*	
Year	*Method*	*Year*	*Research*
		1992	Rejoin OSU
1994	Electronic noses appears in the US	1994 to 1999	Become involved in Electronic nose research to assess factors affecting differentiation
1995	SPME offers advantages over other methods of volatile extraction	1995	SPME explored for evaluation flavor volatiles in whey protein concentrates
1990-1994	Continued improvement in GC equipment and selective detectors	1997	Swiss cheese quality and flavor research becomes a major focus
1995-1999	Chemical basis of flavor beginning to be understood	1999	Begin work with high impact polyfunctional sulfur flavor compounds in cheese

2000-Present

The methods that became available during this decade and those used in my own research program are shown in Table X.

My research program continues to follow the same general direction as it had during the previous decade, with somewhat greater emphasis on Swiss cheese.

The focus of the EN research shifted to try to development an approach to actually determine the basis of the differentiation. This was possible with the Chem Sensor 4400, using mass spectrometry. The first approach was to use normal electronic ionization. We could differentiate volatile headspace volatile compounds from ice cream sticks on the basis of functional groups only (*63-64*). It was shown that differences in carbonyl groups was achieved and that these differences were also related to differentiation by a sensory panel (*65*).

Subsequently, the use of negative chemical ionization was investigated using methane as the ionizing gas. This was successfully used together with descriptive sensory analysis with aged Cheddar cheeses (*66*). Table XI shows the two most abundant masses for selected flavor compounds associated with cheese flavor. The second mass abundance is generally less than 50% of the first.

Table X. Methods introduced and those used my own research program during the decade 2000-present

The World		*My Research*	
Year	*Method*	*Year*	*Research*
2000-present	Analytical methods continue to improve	2002	Electronic nose research enters new phase – developing an NCI/MS library of volatile compounds associated with flavor of milk and cheese
2000 - present	Use of sensory descriptive analysis coupled with GC/MS and model food systems showing promise	2000-present	Swiss cheese research continues
2000 - present	Fourier-transform NIR applied to fatty acids in cheese	2000-2005	Polyfunctional thiols in Cheddar Cheese; Initiate work with FTIR

Table XI. Illustration of NCI MS spectra of selected volatile compounds (unpublished data)

Compound	*Molecular Weight*	*Most abundant masses*	
		First	*Second*
Acetaldehyde	44	59	43
Benzaldehyde	106	121	122
2-Butanone	72	71	72
Diacetyl	86	85	86
Butyric acid	88	87	89

The third and lesser masses (up to 7) make up less than 10% of the first abundance. The primary mass differs with the chemical nature of the compounds. Overall, the most prominent mass is the molecular weight –1 (M-1). In the case of aldehydes the dominant mass in M+15, ketones gives M-1 and diketones give M. A mass library is in the process of being developed to be able to obtain a chemical basis for the differentiation of headspace volatiles by an EN.

Sulfur compounds are well known to be important in the flavor of Cheddar cheese. Previous research has shown that there are multiple sulfur descriptors that are applicable to aged Cheddar cheese (*62*). This suggested that compounds other than those already associated with flavor (methional, hydrogen sulfide, methyl mercaptans, and methyl sulfides) might be present. Investigation revealed that polyfunctional thiols, known to be present in beer, cooked meat and other foods, were also present in aged Cheddar cheese (*66*). Up to 20 compounds could be found in a single cheese. The identification of these compounds is extremely difficult because of the fact that only a few are commercially available and they are present only at ppb levels. Some of the possible compounds found in 11 cheeses, the number of times they were found, the range of concentration and their sensory descriptors are shown in Table XII. Of these only the 4-mercapto-4 methylpentan-2-one compound has been positively identified. The remainder has been tentatively identified based on retention index. Looking at all 11 cheeses there appeared to be up to 40 different possible compounds.

Table XII. Some possible polyfunctional thiols in aged Cheddar cheese (unpublished data)

Compound	*Presence**	*Concentration*	*Descriptor*
4-mercapto, 4 methyl pentan-2-one	10/11	11-43 ppb	Catty, black current
4-mercapto, 4 methyl pentan-2-ol	8/11	1.5-45 ppb	Cooked, sweet
3-mercapto, 3 methyl butanal	7/11	1.5-30 ppb	Broth, cheese, pungent
3-mercapto butanal	3/11	1.5-24 ppb	Broth, cheese, pungent

References

1. Harper, W.J., Swanson, A.M.; Sommer, H.H. *J. Dairy Sci.* **1953**, *26,* 368-372.
2. Harper, W.J.; Swanson, A.M. In *Proceedings of the XIIth International Dairy Congress;* 1949; Section II, Subj 2, pp. 147-155.

3. Harper, W.J. PhD. Thesis 1949. University of Wisconsin, Madison WI.
4. Harper, W.J.; Gould, I.A. *Butter, Cheese and Milk Prods. J.* **1952**, *43(8),* 22-24, 44, 46.
5. Harper, W.J.; Gould, I.A. *Butter, Cheese and Milk Prod. J.* **1952**, *43(9),* 26-30.
6. Harper, W.J.; Maskell, K.T.; Hargrove, R.E. *Butter, Cheese and Milk Prod. J.* **1952**, *54(12),* 20-21, 36-38.
7. Harper, W.J. *J. Dairy Sci.* **1953**, *36,* 808-815.
8. Harper, W.J.; Long, J. E. *J. Dairy Sci.* **1956**, *39,* 129.
9. Long, J.E.; Harper, W. J. *J. Dairy Sci.* **1956**, *39,* 138.
10. Long, J.E.; Harper, W. J. *J. Dairy Sci.* **1956**, *39,* 245.
11. Harper, W.J.; Gould, I.A. *J. Dairy Sci.* **1955**, *38,* 87-95.
12. Harper, W.J. *J. Dairy Sci.* **1955**, *38,* 1391.
13. Harper, W.J. *J. Dairy Sci.* **1957**, *40,* 964.
14. Armstrong, T.V.; Harper, W.J. *Milk Prods. J.* **1954**, *45(5),* 37.
15. Harper, W.J.; Bassett, E.; Gould, I.A. *J. Dairy Sci.* **1954**, *37,* 622-623.
16. Harper, W.J.; Schwartz, D.P.; El-Hagarawy, I.S. *J. Dairy Sci.* **1956**, *39,* 46.
17. Shotwell, T.C.; Harper, W.J.; Gould, I.A. *J. Dairy Sci.* **1953,** *37,* 948-954.
18. Harper, W. J.; Armstrong, T.V. *J. Dairy Sci.* **1954**, *37,* 481-487.
19. Thomas, W.R., Harper, W.J.; Gould, I.A. *J. Dairy Sci.* **1954**, *37,* 717-723.
20. Thomas, W.R., Harper, W.J.; Gould, I.A. *J. Dairy Sci.* **1955**, *38,* 315.
21. El-Hagarawy, I.S.; Slatter, W.L.; Harper, W.J. *J. Dairy Sci.* **1957**, *40,* 579.
22. Handy, M.K.; Harper, W.J.; Weiser, H.H. *J. Dairy Sci.* **1955**, *38,* 147.
23. Hintz, P.C.; Slatter, W.L.; Harper, W.J. *J. Dairy Sci.* **1956**, *39,* 235.
24. Bassett, E.W.; Harper, W.J. In *Proceedings of the XIVth International Dairy Congress;* 1956; Vol. II, Part II, p. 38.
25. Bassett, A.W.; Harper, W.J. *J. Dairy Sci.* **1958**, *41,* 1206.
26. Harper, W.J.; Bassett, E.W. *Milk Products J.* **1958**, *50(8),* 14.
27. Harper, W.J.; Huber, R.M. *J. Dairy Sci.* **1956**, *39,* 1609.
28. Harper, W.J.; Huber, R. *J. Applied Microbiol.* **1961**, *9,* 184.
29. Harper, W.J.; Kristoffersen, T. *J. Dairy Sci.* **1958**, *41,* 998.
30. Hankinson, C.L.; Harper, W.J. *J. Dairy Sci.* **1958**, *41,* 1502.
31. Cole, D.D.; Harper, W.J.; Hankinson, C.L. *J. Dairy Sci.* **1961**, *44,* 141.
32. Harper, W.J.; Gould I.A.; Hankinson, C.L. *J. Dairy Sci.* **1961**, *42,* 2149.
33. Sammuelson, E.G.; Harper, W.J.; Gould, I.A. *Milchwissenschaft* **1961**, *16,* 344.
34. Harper, W.J.; Brown, D.H. *J. Dairy Sci.* **1964**, *47,* 548 (Tech. Note).
35. Pereira, R.R.; Harper, W.J.; Gould, I.A. *J Dairy Sci.* **1966**, *49(11),* 1325.
36. Harper, W.J. In *Proceedings of the XVIII International Dairy Congress;* 1970; Vol. 1E.
37. Harper, W.J.; Gould, I.A. Utilization of radioactive isotopes as tracers in research concerning milk and milk products. International Seminar on Radioisotopes and Radiation, International Atomic Energy Pub. 135, 1966; pp. 15.

38. Pereira, R.R.; Harper, W.J.; Gould I.A. *J. Dairy Sci.* **1968**, *51,* 1850.
39. Periera, R.R.; Harper, W.J.; Gould, I.A. I. Volatile Sulphur Compounds in Heated Milk Products. Centennial Series, Dept. of Dairy Technology, Ohio State University, Pub. No. 7, 1970; pp 1-10.
40. Periera, R.R.; Harper, W.J.; Gould, I.A. II. Factors Affecting the Heat Release of Volatile Sulphur Compounds. Dept. of Dairy Technology, Ohio State University, Centennial Series Pub. No. 7, 1970; pp 11-21.
41. Harper, W.J., Peirera, R.R. In *Proceedings of the XVIII International Dairy Congress;* 1970; Vol. 1E.
42. Harper, W. J. *Milchwissenschaft* **1965**, *20,* 354.
43. Kristoffersen, T.; Harper, W.J. Radioactive Volatile Sulfur Compounds in Cheddar Cheese During Ripening. International Dairy Congress. II. D., 1966; 297.
44. Harper, W.J.; Alfredson, S.; Nillson, R. Acidic and Neutral Carbonyl Compounds in Swedish Cheese. Alnarp Report No. 64. Dec. 1962.
45. Harper, W. J.; Kristoffersen, T. *J. Agr. Food Chem.* **1970**, *18(4),* 563.
46. Harper, W.J.; Carmona, A.; Kristoffersen, T. *J. Food Sci.* **1971**, *36,* 503-506.
47. Harper, W.J.; Wang, J.Y.; Kristoffersen, T. *Milchwissenschaft* **1978**, *33(10),* 604.
48. Lin, Y.C.; Kristoffersen, T.; Harper, W.J. *Milchwissenschaft* **1979**, *34(2),* 69.
49. Harper, W.J.; Wang, J.Y.; Kristoffersen, T. *Milchwissenschaft* **1979**, *34(9),* 525.
50. Harper, W.J.; Carmona de Catril, A.; Chen, J.L. *Milchwissenschaft* **1980**, *36(3),* 129.
51. Harper, W.J.; Wang, J.Y.; Boudreau, D.L. *Milchwissenschaft* **1980**, *35(6),* 340.
52. Harper, W.J.; Wang, J.Y. *Milchwissenschaft* **1980**, *35(9),* 531.
53. Harper, W.J.; Wang, J.Y. *Milchwissenschaft* **1980**, *35(10),* 593.
54. Harper, W.J.; Wang, J.Y. *Milchwissenschaft* **1981**, *36(2),* 70.
55. Harper, W.J.; Blaser, C. *Milchwissenschaft* **1981**, *36(3),* 134.
56. Harper, W.J.; Jin Z. In *Proceedings 3rd International Symposium on Olfaction and Electronic Noses*; Miami, FL, November 4-5, 1996 ; pp 138-148.
57. Harper, W.J.; Sohn, S.; Jou, K.D. In *Proceedings 3rd International Symposium on Olfaction and Electronic Noses*; Miami, FL, November 4-5, 1996; pp 176-184 .
58. Kleinhenz, J.E; Harper, W. J. *Milchwissenschaft* **1997**, *52(4),* 200-204.
59. Yang, M.; Harper, W.J. *Milchwissenscaft* **1988**, *53(4),* 208-212.
60. Harper, W.J. *In Proceedings of the 46th International Congress of Meat Science and Technology;* Buenos Aires, Argentina, August 2000; pp. 282-289.

61. Harper, W.J. *Exp. Med. Biol.* **2001**, *488,* 355-359.
62. Drake, M.A.; Gerard P.D.; Kleinhenze, J.P.; Harper, W.J. *Lebensm. Wissenschaft und Technol.* **2003**, *36,* 13-20.
63. Jiamyangyuen, S.; Harper, W.J. *Milchwissenschaft* **2004**, *59(9/10),* 530-532.
64. Jiamyangyuen, S.; Harper, W.J. 2004. *Milchwissenschaft* **2004**, *59(7/8),* 401-403.
65. Jiamyangyuen, S.; Delwiche, J. F.; Harper, W. J. *J. Dairy Sci.* **2002**, *85(2),* 355-359.
66. Kleinhenze, J.; Kuo, J; Harper. W. J. *Miilchwissenschaft* **2005**, *60,* 275-78.

Chemistry and Analysis

Chapter 2

Flavor Analysis of Dairy Products

Robert J. McGorrin

Department of Food Science and Technology, Oregon State University, 100 Wiegand Hall, Corvallis, OR 97331

The analysis and identifications of volatile flavor compounds in cow's mik, cheeses, and cultured dairy products have been the focus of active flavor research over the past several decades. During that period, the vast preponderance of the published flavor chemistry literature has focused on the qualitative and quantitative composition of volatile components in milk, cheeses, yogurt, and other cultured dairy products. A more contemporary objective of flavor research addresses the flavor relevance of newly identified aroma compounds in dairy-based foods. This chapter provides an overview and perspective of current sample preparation and flavor analysis techniques that are used to identify significant flavor components in dairy food systems. Sensory-directed analytical approaches will be highlighted which have been applied to determine the most important aroma compounds in dairy products and ingredients.

Introduction

The characteristic flavors of milk, cheese, butter, yogurt, sour cream, and other cultured dairy products are generally derived from extremely complex volatile mixtures, often comprising several hundred compounds with diverse chemical structures (*1*). In most instances, the flavor impression of any of these dairy foods is perceived as a single sensation, arising from the sensory integration of a multitude of individual compounds with specific concentrations and ratios (*2*). There is an onging quest among flavor chemists to identify molecules that relate to specific sensory attributes (*3*). The identification of volatile aroma constituents in dairy products often provides unique analytical challenges (*4-5*). A major proportion of this difficulty occurs because key aroma components in dairy products are typically present at significantly lower concentrations as compared with those identified in other foods. In addition, the isolation of dairy flavor compounds is further complicated from difficulties in separation and analytical interferences caused by butterfat. Furthermore, susceptible dairy flavor constituents including mercaptans, sulfides, unsaturated aldehydes and ketones have a tendency to degrade during analysis because of their high reactivities. The goal of modern dairy flavor analysis is to identify key flavor compounds that are produced by dairy fermentation and biogeneration processes. This review will discuss some of the more current volatile isolation and flavor analysis techniques for identification of key flavor components in dairy products.

Motivation for Dairy Flavor Analysis

Volatile components have been identified in dairy products since the mid-1950's, primarily through the development of gas chromatography and its early application to flavor analysis. The flavor chemistry and previous identifications of milk, cheese, and other dairy flavors have been summarized in preceding reviews (*4-8*). Among those reported are the analyses of flavor volatiles in milk (fresh, pasteurized, and UHT-treated), cultured dairy products (butter, buttermilk, sour cream, yogurt) and ripened cheeses (Cheddar, Swiss, Camembert). However, despite these efforts, the flavor significance of these identifications is often unclear. Consequently, it is often impossible to analytically differentiate subtle flavor distinctions in aged Cheddar cheese samples that are perceived as different-tasting. The desired ability to interpret these volatile compound identifications to more effectively control the sensory outcome of dairy flavor formation pathways is often an elusive goal. In parallel with analytical approaches, defined sensory lexicons have been developed to facilitate the linking of specific flavors with the microbial and process technology of Cheddar cheese (*9*). During the past seven years, there has been a resurgence of efforts to develop analytical techniques to characterize key flavors in cow's milk, cultured dairy products and various cheeses. The highlights of these recent developments will be further discussed.

Character-Impact / Key Dairy Flavor Compounds

It has been long been the goal of flavor chemists to elucidate the identity of pure aroma chemicals that have the distinct character-impact of the cheese, fruit, meat, or spice they were derived from. These so-called "character-impact compounds" are unique chemical substances that provide the principal sensory identity or "flavor signature" for a particular food type (*3*). In some instances, character-impact is elicited through a synergistic blend of several aroma compounds. With a few exceptions, a majority of the significant flavor compounds identified to date in dairy products do not individually provide character-impact. The latter is especially true for milk, Cheddar cheese, and cultured products, such as sour cream and yogurt. However, a few examples of character-impact compounds for dairy products are known, including diacetyl (butter), 2-heptanone (blue cheese), and 1-octen-3-ol (Camembert cheese) (*3*). Arguably, butyric acid and propionic acid could also be considered character-impact flavor compounds for Cheddar and Swiss cheeses, respectively. However, while skatole (fecal) and methional (boiled potato) are important constituents of Cheddar cheese flavor (*10-12*), they do not individually provide its unique sensory character (Figure 1). Cheddar flavor is presently understood to arise from a balance of key volatile components, rather than from a unique character-impact compound (*11,12*).

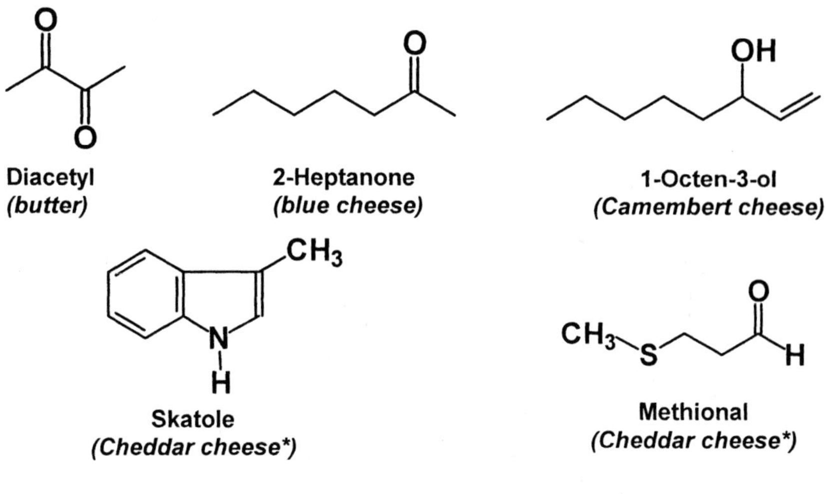

Figure 1. Examples of character-impact and non character-impact (*) dairy flavor compounds.

As recent examples, there have been numerous analytical studies directed towards identifying new key aroma compounds in dairy products. The most significant aroma-active compounds in fresh sour cream butter were elucidated as the character-impact compound diacetyl, with supporting roles from δ-decalactone, (*Z*)-6-dodeceno-γ-lactone, and butyric acid (*13*). Key aroma-active compounds have been reported in dried dairy products including milk and whey powders (*14,15*). Two sensory-guided flavor studies on Cheddar cheese concluded that butyric acid, acetic acid, methional, homofuraneol (5-ethyl-4-hydroxy-2-methyl-(2*H*)-furan-3-one), (*E*)-2-nonenal, (*Z*)-4-heptenal, (*Z*)-1,5-octadien-3-one, and 2-acetyl-1-pyrroline are primary contributors to its pleasant mild flavor (*11,12*). Important contributors to Cheddar aroma are 2,3-butanedione, dimethyl sulfide, dimethyl trisulfide, and methanethiol (*11*). A desirable "nutty" flavor supports "sulfur" and "brothy" characters in a quality aged Cheddar cheese sensory profile. In these recent studies, 2-acetyl-2-thiazoline (*11*) and 2-acetyl-1-pyrroline (*11,12*) contribute "roasted, corny" flavors, which were suggested to be related to the "nutty" Cheddar flavor. In an effort to ascertain the source of the "nutty" flavor in Cheddar cheese, a comprehensive sensory-analytical study concluded that Strecker aldehydes (2-methylpropanal; 2- and 3-methylbutanal), which characteristically are "green, malty, chocolate", provide an enhanced overall "nutty" flavor when added to Cheddar cheese (*16*). A specialty-type British Farmhouse Cheddar cheese was found to contain aroma components *p*-cresol and 2-*iso*-propyl-3-methoxypyrazine, which respectively contribute its characteristic "cowy-barny" and "earthy/soil-like" sensory attributes (*17*). With Cheddar cheese powders, dimethyl sulfide imparts a desirable "creamed corn" flavor (*18*). Representative structures of recently identified dairy flavorants are shown in Figure 2. These include: Furaneol in nonfat dry milk (*14*), sweet whey powder (*15*), whey protein concentrate/isolate (*19*), and heated butter (*20,21*); homofuraneol in mild Cheddar cheese (*11*) and Swiss cheese (*22*); δ-decalactone in butter and buttermilk (*13*); (*Z*)-6-dodecen-γ-lactone in butter (*20*) and mild Cheddar cheese (*11,12*), (*Z*)-4-heptenal in mild Cheddar cheese (*12*), 1-nonen-3-one in milk and yogurt (*23*), 2-acetyl-2-thiazoline (*11,16*) and 2-acetyl-1-pyrroline (*11,12,16*) in mild and aged Cheddar cheese, and 2-*iso*-propyl-3-methoxypyrazine in British Farmhouse Cheddar cheese (*17*).

Dairy Flavor Analysis Techniques

The classical approach to dairy volatile analysis involves a flavor extraction or isolation step, followed by chemical identification (typically using gas chromatography – mass spectrometry) (*24*). The usual protocol for dairy flavor analysis includes sample collection, volatile isolation, concentration, fractionation, separation, and identification. Because of their inherent chemical lability, the proper isolation of dairy flavor samples is critical; mishandling can produce artifacts arising by thermal decomposition, sulfur oxidation, or enzyme-catalyzed reactions (*5*). Prolonged sample heating to increase flavor volatility and recovery can potentially alter the composition of labile dairy components.

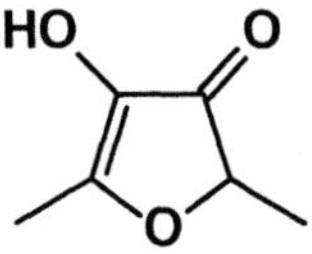

Furaneol
(heated butter, dried milk, sweet whey)

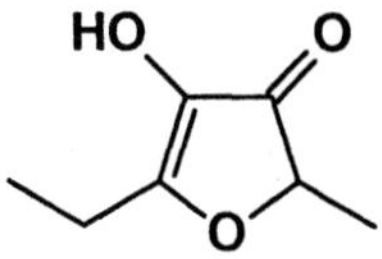

Homofuraneol
(Swiss, Cheddar cheese)

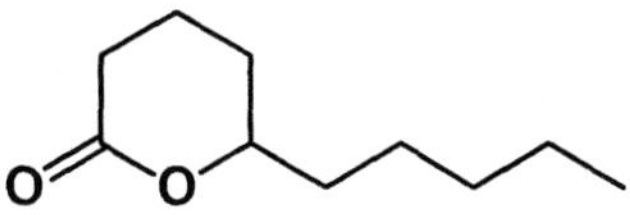

δ -Decalactone
(butter, buttermilk)

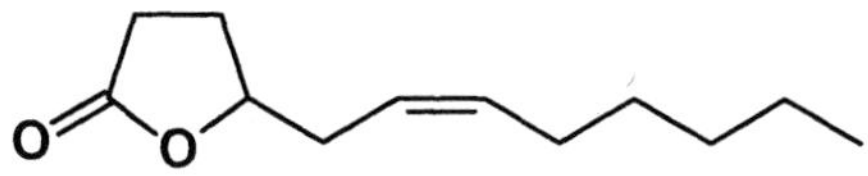

(*Z*)-6-Dodecen-γ-lactone
(Cheddar cheese, butter)

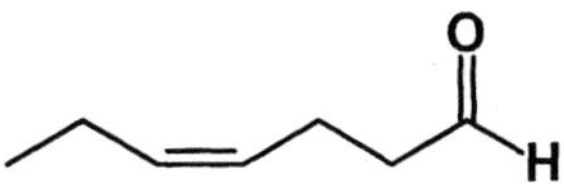

(*Z*)-4-Heptenal
(cream, mild Cheddar)

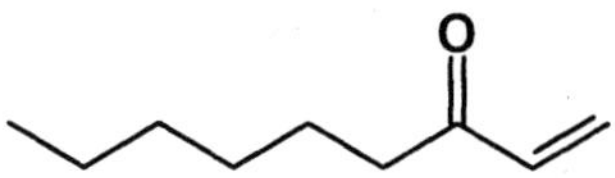

1-Nonen-3-one
(milk, yogurt)

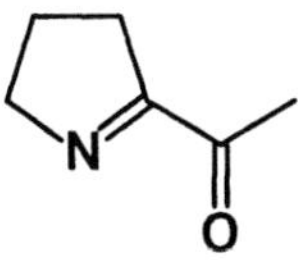

2-Acetyl-1-pyrroline
(Cheddar cheese)

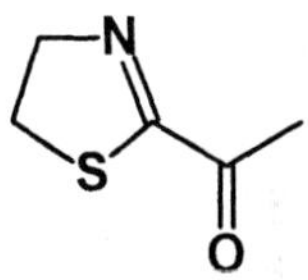

2-Acetyl-2-thiazoline
(Cheddar cheese)

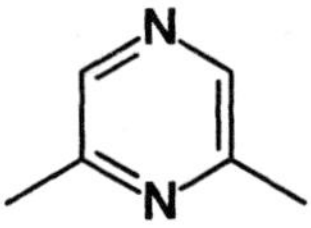

2,6-Dimethylpyrazine
(Parmigiano, dried sweet whey)

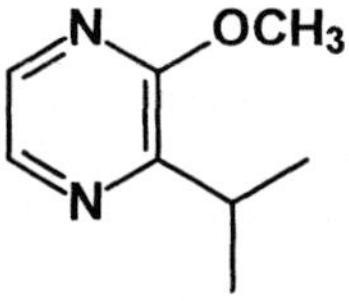

2- *iso*-Propyl-3-methoxypyrazine
(British Farmhouse Cheddar)

Figure 2. Recently identified dairy flavor compounds of sensory significance.

In some cases, sampling choices can influence the analytical outcome; for example, if a cheese sample is obtained at the center core versus close to the surface rind. Similar to other volatile methods, dairy flavor analysis involves a multi-component assay of hundreds of individual flavor compounds, encompassing a wide range of volatilities, polarities, and concentrations (*25*). Several modern methods for isolating and quantifying volatile components in flavors and fragrances have been summarized elsewhere (*26*). The following discussion will focus on those that have been applied to dairy flavor analysis.

High Vacuum Distillation

Due to the difficulties of isolating low levels (ppb to ppt) of volatile components in a fat matrix, classical dairy flavor analyses can be labor and time-intensive. Tedious isolation and concentration processes are often needed to obtain an aroma isolate for quantitative analysis. One of the early classical techniques applied to dairy flavor isolation is vacuum distillation (*27*). Aroma components are distilled from dairy samples for 4 h under high vacuum (~10^{-5} Torr) and mild heat (<60°C) conditions, with subsequent condensation of volatiles in a series of cold traps, followed by solvent extraction. Recoveries can be variable, depending on the relative amount of flavor partitioning between the dairy matrix and the headspace. Losses can also occur during the concentration of the solvent extract. However, the volatile isolate encompasses a broad range of mid- to high-boiling trace-level flavor compounds at sufficient quantities for analysis. This technique has been applied to isolate aroma-active components of Cheddar cheese (*10-12,16*), Chevre-style goat cheese (*28*), nonfat dry milk (*14*), and dried whey protein isolate/concentrate (*19*).

Simultaneous Distillation – Solvent Extraction

Another classic method that has been widely used for flavor volatile isolation is the Likens-Nickerson (simultaneous steam distillation-solvent extraction) technique (*24,29*). When applied to dairy products, this method tends to produce thermal artifacts such as methyl ketones during the concentration and isolation of volatile dairy flavors (*26,30*). A recently improved adaptation of this method is solvent-assisted flavor evaporation (SAFE), in which volatiles are high vacuum-distilled (~10^{-5} Torr) and solvent-extracted in a single step (*31*). Typical SAFE distillation temperatures range from 30-50 °C, significantly reducing the opportunity for thermal artifact formation from labile dairy volatile compounds. A recent model system study compared the analytical recovery yields of the vacuum distillation and SAFE techniques (*32*). A mixture of five flavor compounds, representing a range of volatility and polarity, was dissolved in diethyl ether containing 10% medium-chain triglycerides to simulate a fat-containing food. As shown in Figure 3, the SAFE technique consistently provides higher yields of volatiles than vacuum distillation. This is evident even for polar and less-volatile compounds such as Furaneol (54% recovery), which is typically difficult to volatilize from food systems by vacuum distillation (1% recovery). Examples of the application of this technique include isolation of aroma extracts from dried sweet whey *(15)*, Cheddar cheese *(33)*, Parmesan cheese (26), UHT-milk *(34)*, and milk powder *(35)*.

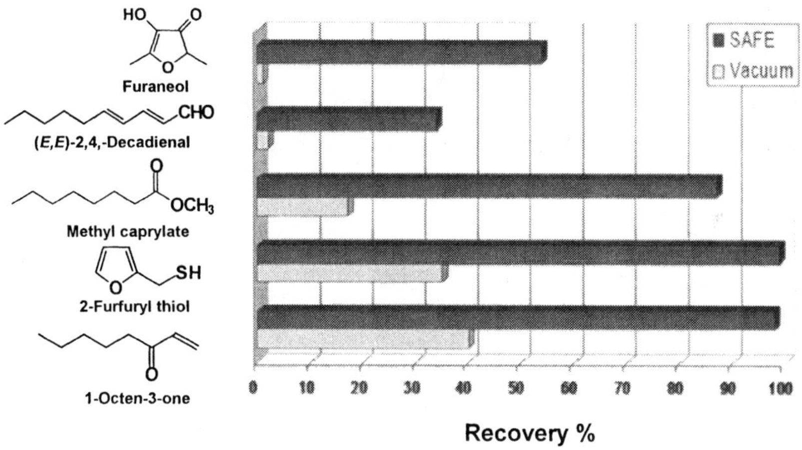

Figure 3. Comparison of volatile recoveries between solvent-assisted flavor evaporation (SAFE) and vacuum distillation in a 10% fat model system. (Adapted from reference 32. Copyright 2001 American Chemical Society.)

Equilibrium Headspace Sampling

Direct headspace sampling techniques avoid issues with fat extraction and chromatographic solvent interferences when applied to dairy flavor analysis. Equilibrium (static) headspace sampling withdraws a 1-mL headspace sample from a septum-capped vial containing the dairy sample. The technique is relatively quick, easily automated and is a useful screening tool, but it lacks the analytical sensitivity needed for detection of low levels of dairy flavor components. If using an autosampler, care must be taken to ensure that samples in the carousel are not undergoing thermally-induced changes when sitting at room temperature for over 2-20 h before the GC analysis. A recent improvement on this technique is solid phase dynamic extraction (SPDE), also known as "the magic needle" *(36)*. In SPDE, headspace volatiles are concentrated onto a 50 mm film of polydimethylsiloxane (PDMS) and 10% activated carbon, which is coated on the inside wall of a stainless steel needle of a 2.5 mL gas tight syringe. The principal advantages of a coated needle are increased sensitivity, with reported 10-fold sensitivity enhancements compared to solid phase microextraction (SPME) using a PDMS fiber. The technique can be easily automated, whereby a fixed volume of dairy headspace sample is withdrawn multiple times (for example, 50 plunger strokes @ 50 ul/sec, each stroke cycle takes 40.5 sec; 50 x 1-mL concentrated headspace volume). The trapped analytes are thermally desorbed from the SPDE needle directly into a heated GC injector with a 1-mL helium volume. A diagram of the SPDE sampling and needle is shown in Figure 4.

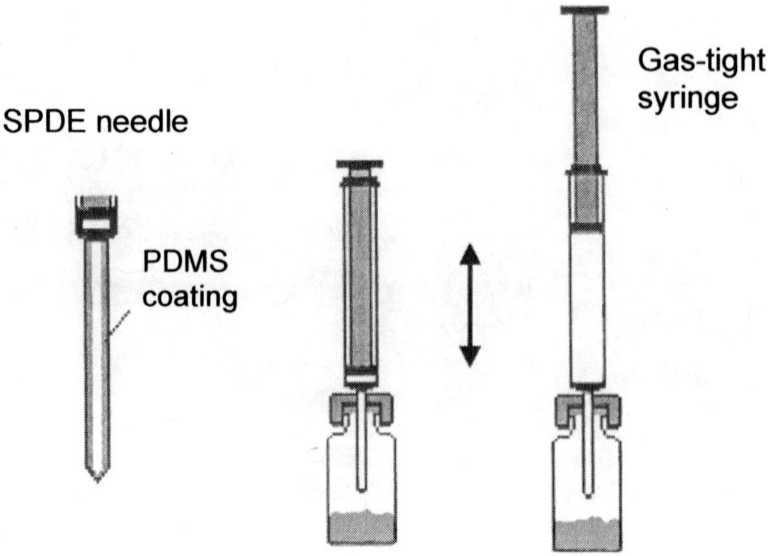

Figure 4. Headspace SPDE sampling device and SPDE needle. (Reproduced from reference 36. Copyright 2004 Elsevier B. V.)

The argument can be made that equilibrium (static) headspace provides a more representative distribution of volatiles in the gas phase, similar to what our olfactory receptors experience during the sensory detection of aromas. Alternatively, dynamic headspace techniques tend to skew volatile concentrates towards higher-boiling compounds, because of the use of high volumes of purge gas. To overcome this problem, recent headspace volatile isolation studies have employed high-volume equilibrium techniques (*21,37*). Figure 5 shows a specially-developed glass headspace concentrator that consists of a water-jacketed 3-liter vessel, containing a chemically-inert Tedlar inflatable bag that can displace up to 1 liter of static headspace through a Tenax trap (*37*). After the dairy sample is allowed to reach equilibrium, the Tedlar bag is gradually inflated with 400 mL of nitrogen over a 20-minute time interval, which slowly displaces the dairy headspace onto the adsorbent trap. At the completion of the concentration, volatiles are desorbed from the trap onto a GC column by the same process used for the dynamic headspace technique. This novel method was recently applied to isolate volatile components from fresh sweet cream butter (*21,37*).

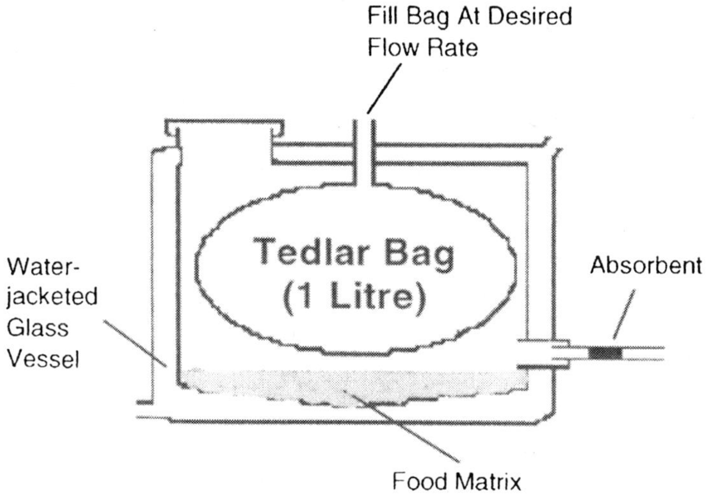

Figure 5. Large volume equilibrium (static) headspace sampling. (Reproduced from reference 37. Copyright 2003 Wiley InterScience.)

Dynamic Headspace Sampling

Dynamic headspace (purge-trap) analysis has often been the mainstay isolation technique in many dairy flavor analysis laboratories, because it provides a balance between speed of analysis and sensitivity of detection (*38-41*). Due to the trapping requirements needed to deliver sufficient analytical sensitivity for dairy flavors, dynamic headspace techniques rely on sweeping high volumes of volatiles through an adsorbent trap packed with Tenax-TA or charcoal (*5*). A combination of adsorbents may provide some selectivity. By selecting Tenax (2,6-diphenyl-*p*-phenylene oxide) as an adsorbent, the primary objective is to efficiently trap and release the desired flavor compounds. An inert gas (helium, nitrogen or argon) is used to displace volatiles from the dairy product onto the trap. Dairy samples are typically purged for 30 min. at 50 °C with a gas flow rate of 100 mL/min onto a glass desorption tube (3.0 mm i.d. x 16 cm length) packed with 100 mg 20/35 mesh of Tenax-TA adsorbent. The trapped flavorants are then thermally desorbed into a GC injection port using carrier gas flow, followed by cryofocusing onto a capillary GC column. Depending on the amount of residual moisture in the trap, it is often "dry purged" with 400 mL of inert gas to remove excess water prior to thermal desorption. Alternatively, graphitized carbon traps have been utilized as adsorbents, which require a microwave thermal desorption unit (Rektorik MW-1A), instead of resistance heating, to displace volatiles from the trap (*39*). Recent examples of the application of the dynamic headspace technique are the isolation of aroma components from Cheddar cheese (*12,16*), Parmigiano Reggiano cheese (*40*), sheep's milk cheese (*39*), and liquid Cheddar whey (*41*).

Solid-Phase Microextraction (SPME)

Solid-phase microextraction (SPME) is a relatively recent technique developed for the isolation and concentration of volatile, semi-volatile, or nonvolatile compounds in foods (*26,42-44*). It is commercially available from Supelco. In dairy flavor applications, it has been typically applied to analyzing volatiles in an equilibrated headspace above a dairy product sample (*45*). SPME integrates volatile sampling, extraction, concentration, and gas chromatographic (GC) injection into a single technique. The method uses a small (1-cm) piece of fused silica coated with a polymeric liquid phase, similar to a GC stationary phase, that absorbs the volatiles and concentrates them on the fiber. The fiber is attached to a stainless steel plunger, sheathed by a protective needle, which is essentially a modified syringe to enable thermal desorption of the analytes into a GC injector port. The selectivity of volatile extraction from the headspace can be significantly altered by the choice of liquid coating on the fiber. With the fiber coating selection, two factors need to be considered: 1) polarity of the flavor analytes, and 2) volatility and molecular weight of the target analytes. Absorbent-type fibers (polydimethylsiloxane, PDMS) partition volatiles into a high-capacity liquid phase, but are most suitable for non-polar analytes. Alternatively, adsorbent-type fibers (PDMS/divinyl benzene (DVB)) trap polar and non-polar volatiles onto porous solids, but have limited capacity. The best compromise for dairy flavor work appears to be the Carboxen/PDMS-coated fiber because of its high capacity, mixed-polarity trapping affinity for flavors and volatile sulfur compounds (*46-49*). However, PDMS/DVB (*50*), and the semi-volatile trapping phases, polyacrylate (*51*) and DVB/Carboxen/PDMS (*52*), have been reported for extracting higher-boiling dairy volatiles in specific applications.

SPME is a deceptively simple technique, and analysts need to be aware of pitfalls that may induce variability in dairy flavor studies where quantitative results are required. Precise control of headspace sampling conditions is critical, since SPME is an equilibrium extraction/desorption process. Because concentrations of dairy analytes are relatively low, attempts to increase volatile recoveries by using higher temperatures can result in competition or displacement of volatile analytes from absorption sites on the SPME fiber. After the sample is allowed to come to equilibration for 15-30 min at 50 °C, the fiber is inserted into the sample headspace. A typical apparatus configuration for volatile concentration using the SPME technique is shown in Figure 6. Factors which especially influence analyte reproducibility and sensitivity during volatile

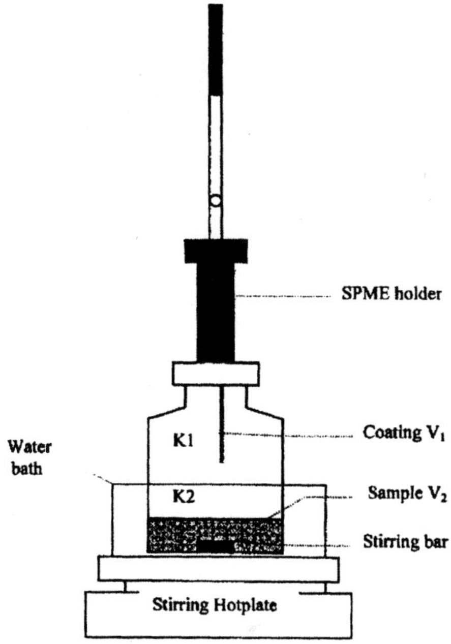

Figure 6. Apparatus for the isolation of headspace volatile flavor compounds on a polymeric-coated SPME fiber (V_1). (Reproduced from reference 53. Copyright 1998 American Chemical Society.)

isolation include dairy sample temperature (40-50 °C), fiber exposure time (5-30 min), and the ratio of sample to headspace volume in the vial. The optimum ratio of sample to headspace depends on the sample's physical state, volatile abundance, and chemical release characteristics; for example, with Parmesan cheese, a 50% sample volume is best, whereas 17% is optimum for orange juice and 66% for soybean oil (*45,53*). Magnetic stirring also plays an important role in reproducibility by accelerating volatile transfer between the sample and the SPME fiber, and shortening the equilibrium time.

An recent application of the SPME technique was evaluated for flavor profiling of various cheese varieties (Cheddar, blue, hard-grated) using 75-μm Carboxen-PDMS fibers (*46*). A comparison of cheese aroma compounds was determined using similar-type cheeses with mild, medium, and aged flavor development. Using a novel sampling approach, the Carboxen-PDMS fibers were exposed to 7g of finely-grated cheese in an equilibrated headspace vial at ambient temperature (22 °C) for a long (16 h) isolation period. Subsequent analysis by gas chromatography-mass spectrometry (GC-MS) and gas chromatography-olfactometry (GC-O) confirmed the presence of compounds previously identified and reported using traditional vacuum distillation or extraction techniques. In typical volatile absorption situations, it would be expected that the concentration of analytes on the fiber would uniformly increase over the 16-h timeframe. To explore the relative efficiency of cheese volatile trapping on the SPME fibers, the authors designed a time-course study to monitor changes in on-fiber volatile concentrates obtained at 1-, 2-, 4-, 9-, and 16-h equilibration times. The relative trapping kinetics for a series of intermediate-volatility odd-chain C_5-C_9 methyl ketones, C_6-C_8 ethyl esters, C_4-C_6 fatty acids, and sulfur compounds were compared. Analyte displacement / competition effects were observed for SPME trapping of blue cheese volatiles, in which high concentrations of 2-heptanone and 2-nonanone occur. Figure 7 shows the relative trapping efficiency of seven blue cheese volatile components

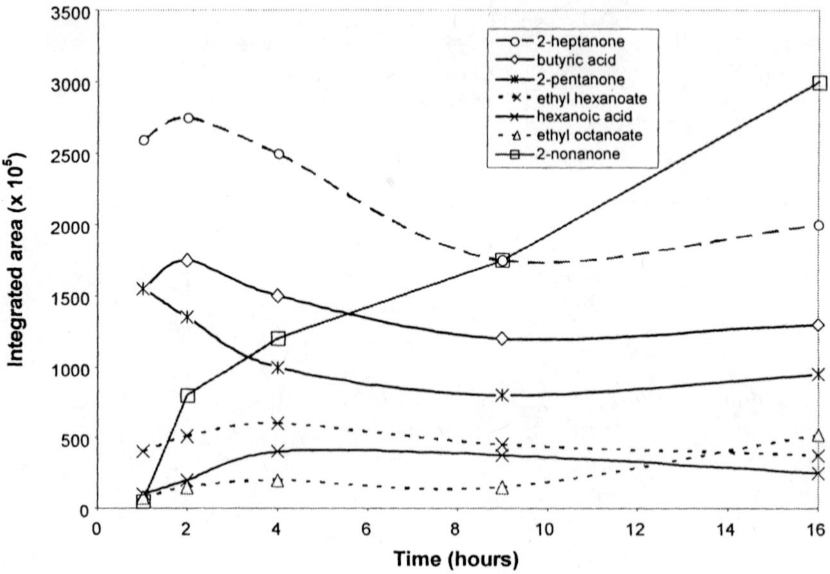

Figure 7. SPME trapping kinetics for representative methyl ketones, ethyl esters, and butyric acid in blue cheese on a Carboxen-PDMS fiber over time (1-16 h). (Adapted from reference 46. Copyright 2003 Elsevier B. V.)

over 16 h. In most cases, a maximum concentration was reached after 8 h; however with longer equilibration times, intermediate-level flavor compounds (e.g., 2-pentanone, 2-hexanone, and 2-heptanone) were displaced by 2-nonanone and ethyl octanoate. Figure 8 displays another comparison from the same study, whereby acetic acid and *iso*-valeric acid reached maximum concentrations on the SPME fiber after 4 h., however displacement effects were noted for 2-heptanol and 3-methylbutanol. In contrast, little or no displacement was observed for trace and medium-concentration cheese aroma components (methyl sulfides, lactones, pyrazines, phenolics). Other cheeses with high levels of a few fatty acids produced similar displacement effects. The noteworthy conclusion was that volatile displacement phenomena need to be considered for so-called "strong-smelling" cheeses with dominant concentrations of a few aromatics such as fatty acids (butyric, hexanoic) in pecorino hard-type cheeses or methyl ketones (2-heptanone) in blue-mold cheeses.

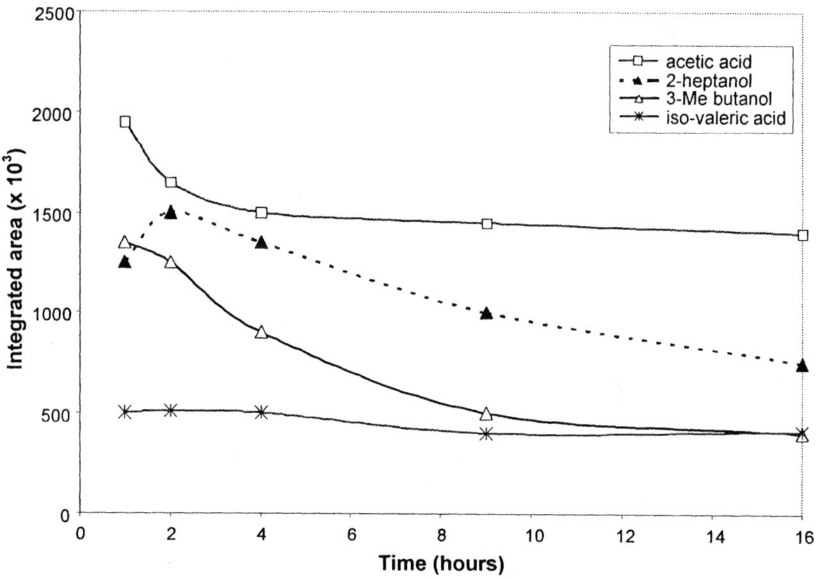

Figure 8. SPME trapping kinetics for representative fatty acids and alcohols in blue cheese on a Carboxen-PDMS fiber over time (1-16 h). (Adapted from reference 46. Copyright 2003 Elsevier B. V.)

Alternatively, SPME was demonstrated to provide excellent quantitation of trace levels of lipid oxidation products related to light-induced off-flavors in milk by GC-MS (*47*). A rapid procedure (15 min., 45 °C) was developed using a 75-μm Carboxen/PDMS SPME fiber to measure pentanal and hexanal in skim and 2% fat milk at 0.5-10 ppb concentrations with a 5-7% coefficient of variance

(CV). Compared to SPME-GC-MS, dynamic headspace-GC-MS afforded poorer sample recoveries and lower precision (8-21% CV). This quality control technique was developed to monitor the development of "metallic" or "cardboard-like" off-flavors in milk, which are typically generated by prolonged exposure to fluorescent light in the supermarket dairy case (*47*).

Other recent applications of SPME are the isolation of aroma components in UHT milk (*52*), artisan goat's milk cheese (*51*), surface-ripened cheese (*49*), Cheddar cheese (*50*), and sulfur components in Cheddar cheese (*48*).

Solid Phase Aroma Concentrate Extraction (SPACE™)

A newly-emerging headspace sampling technique that is a modified version of SPME has been appropriately named SPACE™ (Solid Phase Aroma Concentrate Extraction) (*54*). The method improves on the deficiencies of SPME by utilizing a higher-capacity adsorbent that is coated onto a stainless steel rod (9 cm x 1.2 mm diameter). The dual-adsorbent coating material for SPACE consists of a 9:1 graphite carbon / 2% activated carbon porous polymer film that is doped with Tenax. Following exposure of the SPACE™ rod to the sample headspace for 40 min at 35 °C, analytes are thermally desorbed in a heated GC injector and cryofocused prior to their transfer onto a GC column. Compared to SPME, the volatile isolate obtained by SPACE™ was about 37-times more concentrated, probably due to the trapping ability of the adsorbent, and the 100-times greater surface area of the 9-cm SPACE™ rod versus a 1-cm SPME fiber. The technique should be generally applicable to the concentration of cheese and dairy flavors. A comparison of flavor volatile GC-MS profiles obtained by SPACE™ versus dynamic headspace, SPME, and direct solvent extraction methods for coffee powder is shown in Figure 9. Generally, SPACE™ provided a broader volatile range and increased sensitivity compared with SPME and dynamic headspace sampling. However, further developments of the adsorbent will be required to enable the trapping of high-boiling polar compounds such as Furaneol and related furanones.

Stir Bar Sorptive Extraction (SBSE)

A very recent innovation for analysis of volatile and semi-volatile flavor compounds is stir bar sorptive extraction (SBSE) which has been commercialized by Gerstel under the trade name *Twister* (*55,56*). Flavor volatiles are extracted by stirring the polydimethylsiloxane (PDMS)–coated bar in a liquid sample for a determined time, followed by thermal release of volatiles from the stir bar, either directly into a GC injector or by use of a thermodesorption unit (Gerstel), coupled with capillary GC or GC-MS. Liquid samples can include aqueous food mixtures, polar matrices (acids and alcohols) and fats and oils. The magnetic stir bar is covered with a glass jacket to prevent metal decomposition of the PDMS layer (Figure 10a). Alternatively, when used in the headspace sorptive extraction (HSSE) mode, the stir bar is suspended in a headspace vial over the sample, then thermally desorbed (Figure 10b). A third approach ($SBSE^2$) uses two stir bars for a combination of immersion and headspace sorptive extraction (26). SBSE is based on the same principles as solid-phase microextraction (SPME), except that it has a thicker PDMS coating (25-125 μL) that provides 50 to 250 times greater extraction efficiency (*57*). Consequently, this overcomes the volatile competition and displacement effects that often occur with SPME. However, a disadvantage of SBSE is that it is currently only available with a PDMS coating, and alternate phases would offer enhanced trapping capabilities for a broader range of flavor polarities. Most appropriately, polar phases would be useful for HSSE flavor applications.

In dairy product flavor applications, SBSE has been applied to analyzing flavor compounds in fresh and cooked milk, yogurt, and cream cheese (*58*). Cheese and yogurt samples were diluted 1:1 with water, then transferred to a 10-mL headspace vial and stirred with the SBSE bar for 1 hour at room temperature. The stir bar was rinsed with distilled water, blotted dry, and inserted in thermodesorption unit at 200°C, cryofocused into a glass liner, and transferred via splitless mode onto a Carbowax capillary GC column. A GC-MS total ion chromatogram of a SBSE volatile extract of herb-flavored cream cheese is shown in Figure 11. Cream cheese flavor is primarily culture volatiles such as diacetyl and acetoin plus fatty acids, lactones, and other methyl ketones that provide creamy flavor character in pasteurized milk products (*4,5*). The major identified components were even-carbon C_{10} – C_{16} fatty acids, δ-decalactone and δ-dodecalactone. It is interesting to note that the largest peak in the GC-MS profile is diethylphthalate, which may be an artifact from the packaging material. A pulsed flame photometric detector (PFPD) was used to detect trace sulfur compounds. Volatile evidence that his herb-flavored cream cheese contains garlic is provided by the presence of diallyl di- and trisulfide, and methyl allyl di- and trisulfide. Hydrogen sulfide and dimethyl trisulfide are most likely derived from the butterfat in the cream cheese.

Comprehensive Two-Dimensional Gas Chromatography (GCxGC)

Complex mixtures of volatile dairy flavors are sometimes difficult to completely resolve and analyze using traditional capillary GC-MS. For these situations, a new technique, comprehensive two-dimensional gas chromatography (GCxGC), was recently developed (*59,60*). Similar to traditional multidimensional GC (GC-GC), this instrumental method uses two GC columns of different polarity that are linked in series, and analytes are switched from the first to the second column to improve the separation power. However, the unique feature of GCxGC is that both columns are directly

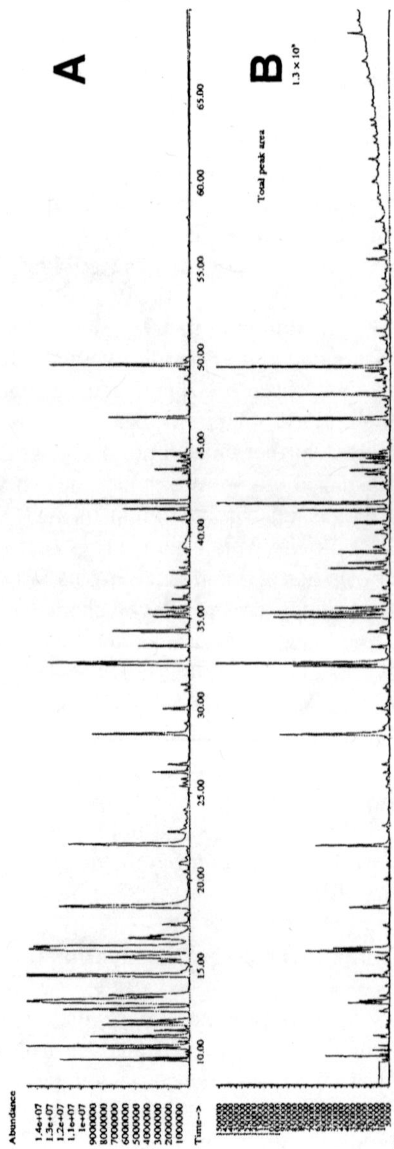
A
B
Abundance
Time-->
Total peak area
10.00
15.00
20.00
25.00
30.00
35.00
40.00
45.00
50.00
55.00
60.00
65.00

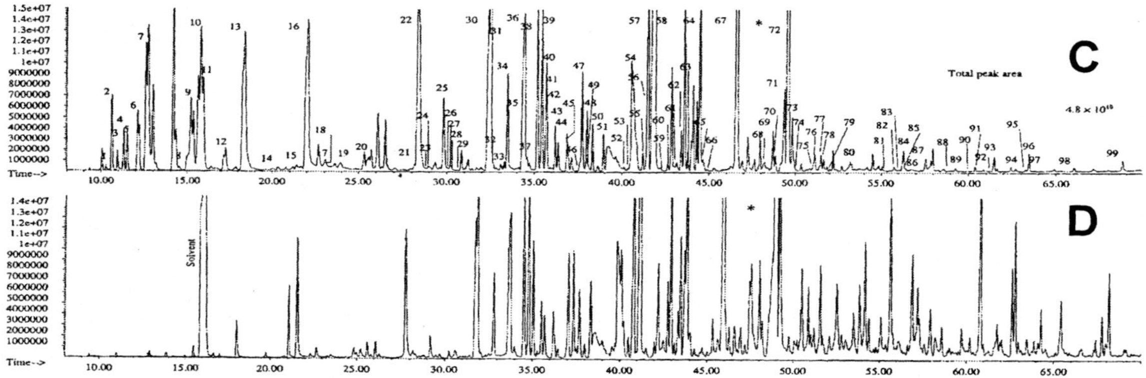

Figure 9. Comparison of GC-MS total ion chromatograms of coffee powder obtained by four volatile concentration techniques: A) dynamic headspace; B) SPME (100-µm PDMS); C) SPACE™; D) solvent extraction (CH_2Cl_2). (Reproduced from reference 54. Copyright 2004 Wiley InterScience.)

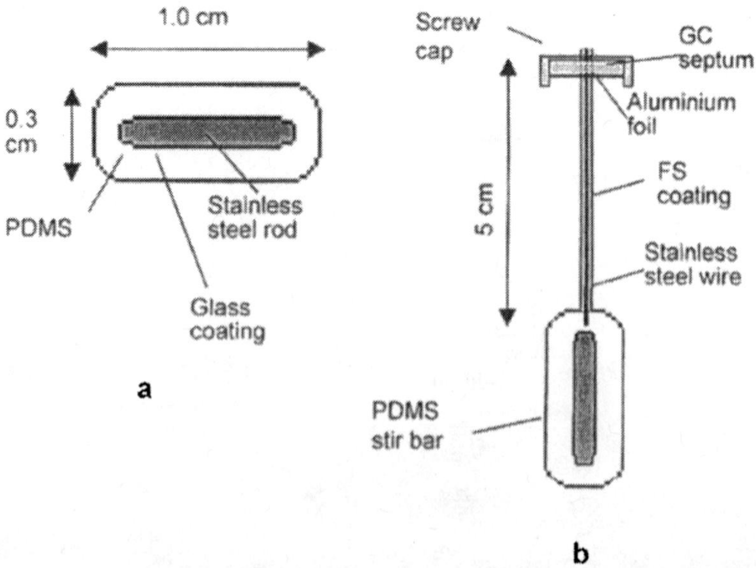

Figure 10. Elements of stir bar sorptive extraction: a) SBSE stir bar coated with PDMS; b) alternative configuration for HSSE mode; FS = fused silica. (Reproduced from reference 57. Copyright 2002 American Chemical Society.)

connected to each other in the same GC oven, and there is no switching valve. Normally, this would produce very broad peaks on the second column, and not much enhancement in the separation power of the dual system. The solution is provided by a specially designed longitudinally-modulated cryogenic trap at the GC column interface, which regulates and focuses the transfer of analytes between columns (*59*). The trap consists of a 3-cm hollow CO_2-cooled sleeve that slides along the outside of the capillary column interface. During the GCxGC separation, the cryotrap is programmed to oscillate back-and-forth every 5 seconds, such that packets of analytes are zone-compressed, then automatically diverted or "pulsed" to a second GC column with a different type of separation phase. A GCxGC instrumental configuration is shown in Figure 12.

The detector for the GCxGC system is time-of-flight mass spectrometry (TOF-MS), which is required because of the higher spectral acquisition rate (500 spectra/sec), compared to quadrupole-MS or ion-trap mass analyzers. Because TOF-MS does not involve the use of a scanning field for mass measurement, it offers significantly improved MS acquisition speed sensitivity. A computerized data system is required to plot the 2-D spectral output.

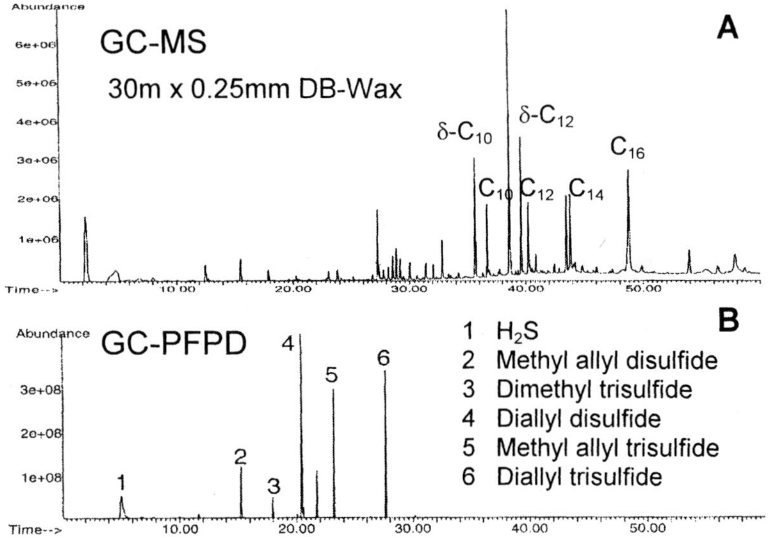

Figure 11. Comparison of SBSE extract chromatograms of herb cream cheese obtained by: A)GC-mass spectrometry B)GC-pulsed flame photometric detection (Reproduced from reference 58. Copyright 2000 Gerstel.)

An application of GCxGC analysis for a complex dairy flavor mixture is provided for the aroma constituents of sour cream (*61*). A volatile concentrate was afforded using the SAFE simultaneous distillation-extraction technique. The GC column configuration consisted of a non-polar (CP-Sil 5, dimethylpolysiloxane) pre-column, followed by a shorter, more polar (BPX-50, 50% phenylpolysiloxane) analytical column. Since the second (analytical) column is eluting about 6-times faster than the pre-column, the GC separation of each volatile transfer "packet" is completed before the next one begins to elute. A 2-D MS computer-generated spectral array plots the total ion chromatogram from the first column (no GCxGC) versus the transfer ion chromatogram output from the second column. At any given position in the total ion chromatogram (x-axis), a corresponding higher-resolved slice is shown for the multidimensional GCxGC separation on the second column (y-axis; see Figure 13).

A further example of the applicability of GCxGC in a dairy-based spread was the difficult chromatographic separation and MS identifications of two pairs of co-eluting flavor components: 2-heptanone with methional, and 2-nonanone with sotolon. For one of the pairs, the GCxGC TOF-MS technique was utilized to resolve the minor component, sotolon, from the major interference peak, 2-

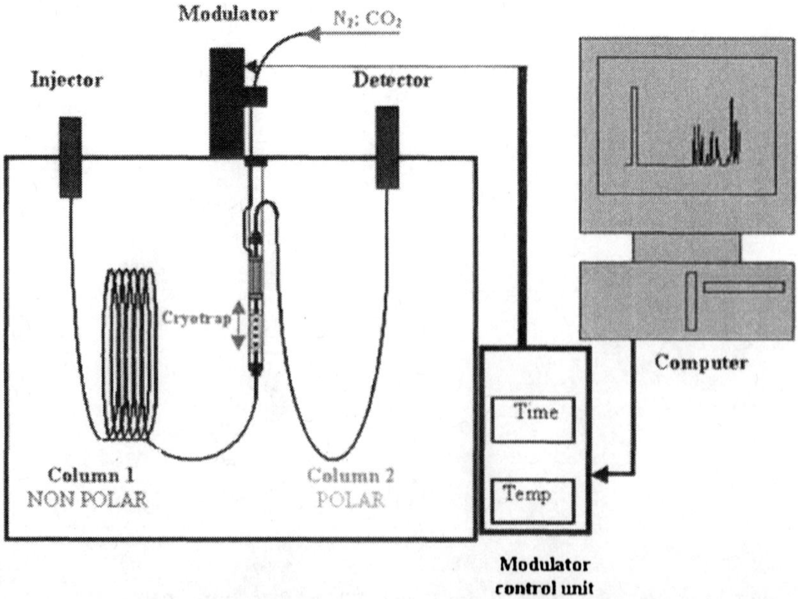

Figure 12. Longitudinally-modulated comprehensive two-dimensional gas chromatography (GCxGC). (Reproduced from reference 59. Copyright 2001 AOAC Intl.)

nonanone. The outcome is presented in Figure 14. The left cluster of peaks is the output from the first column, and the right peak cluster is produced after 2-D separation on the second column. The solid trace is the total ion chromatogram, and the dashed trace is the selected ion chromatogram for the 128 ion. The molecular ion of sotolon is 128, however the major peak, 2-nonanone also has a 128 fragment ion. After the second separation, the amount of the interfering compound is greatly reduced in the mass spectrum, allowing an easy confirmation of the correct structural assignment to be made for sotolon.

Sensory-Directed Analytical Techniques

Progress in the instrumental analysis of cheese and dairy flavors has often provided long lists of volatiles. Until recently, the sensory relevance of these volatile compounds has not been as extensively evaluated. Prior to the late 1990's, much has been published on dairy flavors, often without identifying the

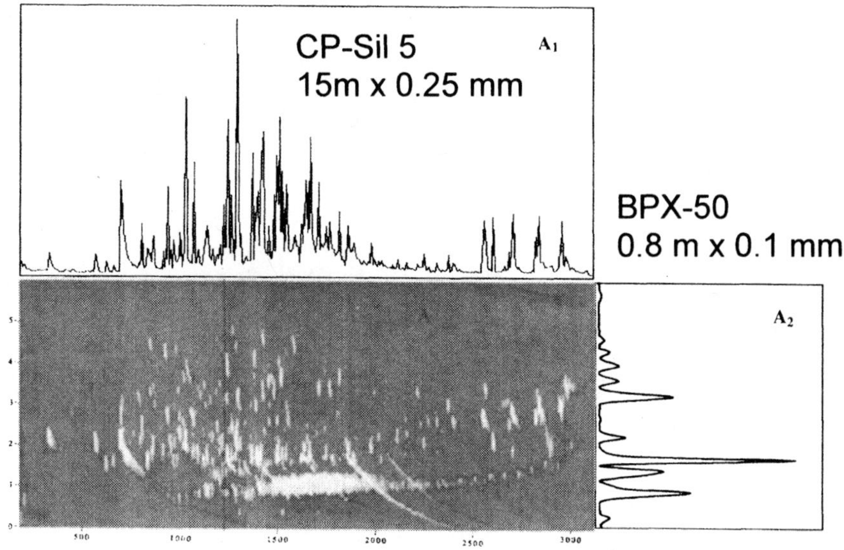

Figure 13. GCxGC TOF-MS analysis of sour cream flavor extract. (Reproduced from reference 61. Copyright 2003 J Chrom.)

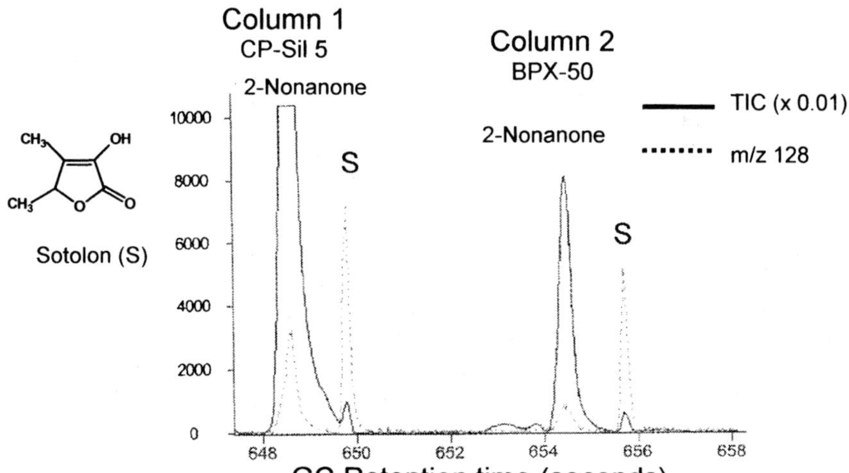

Figure 14. GCxGC TOF-MS resolution of 2-nonanone from sotolon in dairy spread. (Reproduced from reference 61. Copyright 2003 J Chrom.)

impact compounds. Therefore, one of the major objectives in flavor research is to select those compounds that significantly contribute to the characteristic aroma.

GC-Olfactometry (GC-O) Techniques

In general, the aroma of dairy flavors consists of many volatile compounds, only a few of which are sensorially relevant. The first essential step in aroma analysis is the differentiation of the more potent volatile odorants from those having low or no aroma activity. Several approaches have been taken to estimate the sensory contribution of single odorants to the overall aroma of a food. The application of sensory-directed analytical flavor techniques for evaluation of key aroma compounds in cheese and dairy products has been previously reviewed (*4,5*). Dilution techniques and time-intensity measurements are the two principal GC-olfactometry (GC-O) methods.

Odor Dilution: Two principal techniques have been developed: CharmAnalysis™ (Combined Hedonic Aroma Response Measurements) (*62*) and AEDA (Aroma Extract Dilution Analysis) (*63*). Both methods use serial dilutions of the volatile fraction, followed by GC-O evaluation after each successive dilution. The process is repeated until no detectable odor is perceivable in the GC effluent. In the AEDA technique, the highest dilution at which an aroma compound can be smelled is defined as it flavor dilution (FD) factor. While the duration of the odor is also factored into the Charm Analysis, the peak maximum of a CharmAnalysis value is identical to the FD factor. Aroma Extract Concentration Analysis (AECA) (*63*) involves an aroma extract enrichment process by successive solvent distillation and GC-O steps.

Time Intensity: The Osme technique measures the perceived odor intensity of a compound in the GC effluent (*64*). This method generates an FID-style chromatogram (an "Osmegram"). Ideally, it requires only one injection when used by well-trained flavor assessors. The flavorist rates the aroma intensity utilizing a computerized 16-point scale time-intensity device, and indicates the corresponding aroma characteristics and changes in odorant concentration. Significant aroma compounds were evaluated in Parmigiano Reggiano cheese (*40*) and whey powder (*15*) by a combination of Osme and AEDA methods.

Odor/flavor Threshold Index: Sensory techniques that measure the relative potency of individual flavor compounds have been developed and used in conjunction with GC-O methods. These include odor unit (based on nasal odor thresholds), flavor unit (from retronasal odor thresholds), and odor activity value (OAV; the ratio of an odorant's concentration to its threshold concentration). However, these concepts first require identification and quantification of volatile compounds, then determination of their unique threshold values. Recent studies suggest that OAV's may not be good indicators of an odorant's relative contribution to the overall intensity of a flavor mixture, and other flavor volatile assessment methods may require further development (*65*).

A contemporary sensory-guided dairy aroma research program combines sensory and analytical approaches (*66,67*). It is principally composed of the following four key steps, which can be applied to the characterization of both cheese and dairy aromas and off flavors:

1) Qualitative aroma composition of the most potent aroma components by Osme, AEDA/AECA, or CharmAnalysis™ (based on GC-Olfactometry)
2) Quantitative aroma composition of essential aroma constituents using isotopically-labeled standards (isotope dilution assay)
3) Determination of odor thresholds, and calculation of odor activity values (odor activity value concept)
4) Confirmation of analytical data by sensory assessment of recombinated mixtures (aroma simulation based on quantitative data)

The last step is essential and validates the analytical results. However a crucial factor is accurate quantification of the aroma-impact compounds. A limitation of such dilution techniques is that they often do not account for losses of highly volatile compounds during the extraction and concentration process.

In conclusion, there is a clear need to improve the quality and stability of cheese and dairy aromas and flavors. The techniques presented in this chapter represent realistic and practical approaches for analyzing aromas more purposefully. Depending on the application, it is possible to simplify the approach and to find compromises in the dairy flavor isolation, separation, and sensory assessment stages, so that essential results can be obtained in a reasonable time. The food and flavor industry is well advised to profit from this development and adapt these different techniques to their specific needs.

Literature Cited

1. Nijssen, L. M.; Visscher, C. A.; Maarse, H.; Willemsens, L. C.; Boelens, M. H., Eds., *Volatile Compounds in Food. Qualitative and Quantitative Data*, 7th ed. TNO Biotechnology and Chemistry Institute, Zeist, The Netherlands, 1996.
2. Reineccius, G. *Flavor Chemistry and Technology*, 2nd ed., CRC Press / Taylor & Francis, Boca Raton, FL, 2006.
3. a) McGorrin, R. J. in *Flavor, Fragrance, and Odor Analysis*, Marsili, R. ed., Marcel Dekker, New York, NY, 2002, pp. 375-413; b) McGorrin, R. J. in *Sensory-Directed Flavor Analysis*, R. Marsili, Ed., Taylor & Francis, Boca Raton, FL, 2006, pp 223-268.
4. Parliment, T. H.; McGorrin, R. J. in: *Flavor Chemistry: Industrial and Academic Research*, Risch; S. J.; Ho, C.-T., Eds., ACS Symposium Series 756, American Chemical Society, Washington, D.C., 2000, pp 44-71.

5. McGorrin, R. J. in *Food Flavors and Chemistry: Advances of the New Millennium*, Spanier, A. M.; Shahidi, F.; Parliment, T. H.; Mussinan, C. J.; Ho, C.-T.; Contis E. T., Eds., Royal Society of Chemistry, Cambridge, 2001, pp 67-84.
6. Nursten, H. E., *Int. J. Dairy Technol.* **1997**, *50*, 48-56.
7. Urbach, G *Int. J. Dairy Technol.* **1997**, *50*, 79-89.
8. Singh, T. K.; Drake, M. A.; Cadwallader, K. R., *Compr. Rev. in Food Science and Food Safety*, **2003**, *2*, 139-162.
9. Drake, M. A. *J. Dairy Sci.*, **2004**, *87*, 777-784.
10. Christensen, K. A.; Reineccius, G. A. *J. Food Sci.* **1995**, *60*, 218-220.
11. Milo, C.; Reineccius, G. A. *J. Agric. Food Chem.* **1997**, *45*, 3590-3594.
12. Zehentbauer, G.; Reineccius, G. A. *Flavour Fragr. J.* **2002**, *17*, 300-305.
13. Schieberle, P.; Gassenmeier, K.; Guth, H.; Sen, A.; Grosch, W. *Lebensm. Wiss. und Technol.* **1993**, *26*, 347.
14. a) Karagul-Yuceer, Y.; Drake, M. A.; Cadwallader, K. R. *J. Agric. Food Chem.* **2001**, *49*, 2948-2953; b) Karagul-Yuceer, Y.; Drake, M. A.; Cadwallader, K. R. in *Freshness and Shelf Life of Foods*, Cadwallader, K.R., Weenen, H., Eds., ACS Symposium Series 836, American Chemical Society, Washington, D.C., 2003, pp 108-123.
15. Mahajan, S. S.; Goddik, L.; Qian, M. C. *J. Dairy Sci.* **2004**, *87*, 4057-4063.
16. Avsar, Y. K.; Karagul-Yuceer, Y.; Drake, M. A.; Singh, T. K.; Yoon, Y.; Cadwallader, K. R. *J. Dairy. Sci.* **2004**, *87*, 1999-2010.
17. Suriyaphan, O.; Drake, M. A.; Chen, X. Q.; Cadwallader, K. R. *J. Agric. Food Chem.* **2001**, *49*, 1382-7.
18. Marsili, R. Application of SPME GC-MS for flavor analysis of cheese-based products, American Chemical Society, Philadelphia, PA, August 25, 2004.
19. Carunchia Whetstine, M. E; Croissant, A. E.; Drake, M. A. *J. Dairy Sci.*. **2005**, *88*, 3826.
20. Budin, J. T.; Milo, C.; Reineccius, G. A. in: *Food Flavors and Chemistry: Advances of the New Millennium*, Spanier, A. M.; Shahidi, F.; Parliment, T. H.; Mussinan, C. J.; Ho, C.-T.; Contis, E. T., eds., Royal Society of Chemistry, Cambridge, 2001, pp 85-96.
21. Peterson, D. G.; Reineccius, G. A. *Flavour Frag. J.* **2003**, *18*, 320-324.
22. Preininger, M.; Grosch, W. *Lebensm. Wiss. und Technol.* **1994**, *27*, 237.
23. Ott, A.; Fay, L. B.; Chaintreau, A., *J. Agric. Food Chem.* **1997**, *45*, 850-858.
24. Teranishi, R.; Kint, S. in Flavor *Science: Sensory Principles and Techniques*, Acree, T. E.; Teranishi, R., Eds., ACS Professional Reference Book, American Chemical Society, Washington, DC, 1993, pp 137-167.
25. Mussinan, C. J.; Morello, M. J., Eds. *Flavor Analysis: Developments in Isolation and Chacterization*, ACS Symposium Series 705, American Chemical Society, Washington, D.C., 1998.

26. Werkhoff, P.; Brennecke, S.; Bretschneider, W.; Bertram, H.-J. in *Flavor, Fragrance, and Odor Analysis*, Marsili, R. ed., Marcel Dekker, New York, NY, 2002, pp 139-204.
27. Sen, A.; Laskawy, G; Schieberle, P.; Grosch, W. *J. Agric. Food Chem.* **1991**, *39*, 757-759.
28. Carunchia Whetstine, M. E.; Karagul-Yuceer, Y.; Avsar, Y. K.; Drake, M. A. *J. Food Sci.* **2003**, *68*, 2441.
29. Parliment, T. in *Flavor, Fragrance, and Odor Analysis*, Marsili, R. ed., Marcel Dekker, New York, NY, 2002, pp 1-23.
30. Dirinck, P.; Winne, A. *J. Chromatogr. A.* **1999**, *847*, 203-208.
31. Engel, W.; Bahr, W.; Schieberle, P. *Eur. Food Res. Technol.* **1999**, *209*, 237-241.
32. Blank, I.; Devaud, S.; Fay, L. B.; Cerny, C.; Steiner, M.; Zurbriggen, B. in *Aroma Active Compounds in Foods: Chemistry and Sensory Properties*, Takeoka, G. R.; Guntert, M.; Engel, K.-H., Eds., ACS Symposium Series 794, American Chemical Society, Washington, D.C., 2001, pp 9-20.
33. Carunchia-Whetstine, M. E.; Cadwallader, K. R.; Drake, M. A. *J. Agric. Food Chem.* **2005**, *53*, 3126-3132.
34. Colahan-Sederstrom, P. M.; Peterson, D. G. *J. Agric. Food Chem.* **2005**, *53*, 398-402.
35. Preininger, M; Ulrich, F. in *Gas Chromatography-Olfactometry: The State of the Art*, Leland, J. V.; Schieberle, P.; Buettner, A.; Acree, T. E., Eds., ACS Symposium Series 782, American Chemical Society, Washington, D.C., 2001, pp 46-61.
36. Bicchi, C.; Cordero, C.; Liberto, E.; Rubiolo, P.; Sgorbini, B. *J. Chromatogr. A.* **2004**, *1024*, 217-226.
37. Peterson, D. G; Reineccius, G. A. *Flavour Frag. J.* **2003**, *18*, 215-220.
38. Ott, A.; Germond, J. E.; Baumgartner, M.; Chaintreau, A. *J. Agric Food Chem.* **1999**, *47*, 2379-2385.
39. Barron, L. J. R.; Redondo, Y.; Ortigosa, M.; Carbonell, M.; Garcia, S.; Torre, P.; de Renobales, M.; Fernandez-Garcia, E. *Lait*, **2005**, *85*, 491-513.
40. Qian, M.; Reineccius, G. *Flavor Fragr. J.* **2003**, *18*, 252-259.
41. Carunchia Whetstine, M. E.; Parker, J. D.; Drake, M. A.; Larick, D. K. *J. Dairy Sci.* **2003**, *86*, 439-448.
42. Pawliszyn, J. *Solid Phase Microextraction: Theory and Practice*, Wiley-VCH, New York, NY, 1997.
43. Wercinski, S. A. S., Ed., *Solid Phase Microextraction: A Practical Guide,* Marcel Dekker, New York, NY, 1999.
44. a) Harmon, A. D. in *Flavor, Fragrance, and Odor Analysis*, Marsili, R. ed., Marcel Dekker, New York, NY, 2002, pp. 75-106; b) Marsili, R., *ibid*, pp 205-227.
45. Lee, J.-H.; Diono, R.; Kim, G.-Y.; Min, D. B. *J. Agric. Food Chem.* **2003**, *51,* 1136-1140.

46. Frank, D. C.; Owen, C. M.; Patterson, J. *Lebensm. Wiss. und Technol.* **2004**, *37*, 139-154.
47. Marsili, R. *J. Chromatogr. Sci.* **1999**, *37*, 17-23.
48. Burbank, H. M.; Qian, M. C. *J. Chromatogr. A.* **2005**, *1066*, 149-157.
49. Lecanu, L.; Ducruet, V.; Jouquand, C.; Gratadoux, J. J.; Feigenbaum, A., *J. Agric. Food Chem.* **2002**, *50,* 3810-3817.
50. Dufour, J.-P.; Delbecq, P.; Albela, L. P. in *Gas Chromatography-Olfactometry: The State of the Art*, Leland, J. V.; Schieberle, P.; Buettner, A.; Acree, T. E., Eds., ACS Symposium Series 782, American Chemical Society, Washington, D.C., 2001, pp 123-137.
51. Guillen, M. D.; Ibargoita, M. L.; Sopelana, P.; Palencia, G.; Fresno, M. *J. Dairy Sci.* **2004**, *87*, 284-289.
52. Vazquez-Landaverde, P. A.; Velazquez, G.; Torres, J. A.; Qian, M. C. *J. Dairy Sci.* **2005**, *88*, 3764-3772.
53. Jia, M.; Zhang; H.; Min, D. B. *J. Agric. Food Chem.* **1998**, *46,* 2744-2747.
54. Ishikawa, M.; Ito, O.; Ishizaki, S.; Kurobayashi, Y.; Fujita, A. *Flavor Fragr. J.* **2004**, *19*, 183-187.
55. Baltussen, E.; Sandra, P.; David, F.; Cramers, C. *J. Microcolumn Sep.* **1999**, *11*, 737-747.
56. Majors, R. E. *LCGC North America*, **2003**, *21*, 108-118.
57. Bicchi, C.; Iori, C.; Rubiolo, P.; Sandra, P. *J. Agric Food Chem.* **2002**, *50*, 449-459.
58. Hoffman, A.; Heiden, A; Determination of flavor and off-flavor compounds in dairy products using stir bar sorptive extraction (SBSE) and thermal desorption GC/MSD/PFPD, paper D34, *Proc. 23rd intl. symp on Capillary Chrom.* (IOPMS, Kortrijk, Belgium, 2000. Application note 5-2000. Gerstel. www.gerstel.com
59. Truong, T. T.; Marriott, P. J.; Porter, N. A. *J. AOAC Intl.* **2001**, *84*, 323-335.
60. Marriott, P. and Shellie, R. *Trends in Analytical Chemistry*, **2002**, *21*, 573-583.
61. Adahchour, M.; van Stee, L. L. P.; Beens, J.; Vreuls, R. J. J.; Batenburg, M. A.; Brinkman, U. A. Th. *J. Chromatogr. A* **2003**, *1019*, 157-172.
62. Mayol, A. R.; Acree, T. E. in *Gas Chromatography-Olfactometry: The State of the Art*, Leland, J. V.; Schieberle, P.; Buettner, A.; Acree, T. E., Eds., ACS Symposium Series 782, American Chemical Society, Washington, D.C., 2001, pp 1-10.
63. Grosch, W.; Kerscher, R.; Kubickova, J.; Jagella, T. *ibid*, p. 138-147.
64. McDaniel, M. R.; Miranda-Lopez, R.; Watson, B. T.; Micheals, N. J.; Libbey, L. M. in *Flavors and Off-flavors*, Charalambous, G., Ed., Elsevier, Amsterdam, 1990, pp 23.
65. Audouin, V.; Bonnet, F.; Vickers, Z.; Reineccius, G. A. in *Gas Chromatography-Olfactometry: The State of the Art*, Leland, J. V.;

Schieberle, P.; Buettner, A.; Acree, T. E., Eds., ACS Symposium Series 782, American Chemical Society, Washington, D.C., 2001, pp 156-171.
66. Engel, K.-H. in *Aroma Active Compounds in Foods: Chemistry and Sensory Properties*, Takeoka, G. R.; Guntert, M.; Engel, K.-H., Eds., ACS Symposium Series 794, American Chemical Society, Washington, D.C., 2001, pp 1-5.
67. Blank, I. in *Flavor, Fragrance, and Odor Analysis*, Marsili, R. ed., Marcel Dekker, New York, NY, 2002, pp 297-331.

Chapter 3

Establishing Links between Sensory and Instrumental Analyses of Dairy Flavors: Example Cheddar Cheese

MaryAnne Drake[1], Keith R. Cadwallader[2], and Mary E. Carunchia Whetstine[1]

[1]Department of Food Science, Southeast Dairy Foods Research Center, North Carolina State University, Campus Box 7624, Raleigh, NC 27695
[2]Department of Food Science and Human Nutrition, University of Illinois, 1302 West Pennsylvania Avenue, Urbana, IL 61801

The flavor of dairy products is an important aspect of their overall appeal. Understanding flavor is crucial to effective and strategic research and marketing. Part of understanding flavor involves linking sensory flavor attributes with the volatile chemical components responsible for their occurrence. Establishing these relationships can be challenging and requires the combination of comprehensive sensory descriptive and instrumental analyses. Specific examples of identifying chemical compounds responsible for sensory-perceived flavors (rosy/floral, nutty, mothball/grassy) in Cheddar cheese are presented.

The flavor of Cheddar cheese is a key parameter for consumer acceptance and marketing (*1*). Using combined instrumental-sensory analysis, many sensory and chemical aspects of Cheddar flavor have been characterized (*1, 2*). When specific links are definitely established between volatile flavor compounds and sensory attributes of cheese, an enhanced understanding of cheese flavor is achieved and powerful information for linking flavor to production technology is obtained (*3, 4*). Such studies require the correlation of chemical and descriptive sensory analysis results.

Instrumental techniques for volatile compound extraction and identification are reviewed elsewhere (*2, 5*). It is important to note that any volatile compound extraction approach has strengths as well as shortcomings. No single approach will extract all volatile components with equal recovery so a combination of techniques (e.g. headspace and solvent extraction) should be used if a complete picture of volatile compounds is desired (*2, 5*). Extracted volatile components should be identified as a starting point, but additionally, compounds contributing to flavor should also be characterized using gas chromatography olfactometry (GC-O) (*2, 6*). GC-O can accomplish multiple things including tentative identification of compounds, elimination of compounds without odor activity from the list of volatile compounds, and selection of key aroma compounds (based on intensities and odor characteristics). Further, the human nose may be more sensitive to aroma compounds than the most sensitive instrumental detector, and GC-O allows those compounds present at concentrations below instrumental detection limits to be characterized.

Descriptive sensory analysis is often, and unfortunately, overlooked in flavor chemistry studies. Flavor is a sensory perception, and without sensory analysis, the best instrumental analysis has no direct relevance to flavor characterization. When linking results of instrumental analysis to perceived aroma, descriptive sensory analysis is the appropriate sensory tool to use (*7*) since this type of sensory tool provides objective, reproducible, and analytical information on the sensory perception of the product flavor – analogous to instrumental data. Descriptive sensory analysis consists of a trained panel where individuals function in unison as components of a "sensory instrument" to document and describe the sensory attributes of a product (*7*). The panel uses a lexicon (language or group of words) to describe the sensory properties of the product. The lexicon has definitions and references for each sensory attribute such that panel results can be reproduced and facilitates correlation with instrumental data (*7*).

The general approach to linking sensory perception with analytical data involves three basic steps: 1) selection of products with the desired or target flavors using descriptive sensory analysis, 2) instrumental analysis of volatile compounds, and 3) confirmation of the sensory role of the volatile compounds using quantitation, threshold analysis, and descriptive sensory analysis of model systems. When both instrumental and sensory analysis techniques are used optimally, the results can be powerful as concrete relationships between sensory perception and volatile components can be established. This paper describes three examples of the application of these approaches to Cheddar cheese flavors: rosy/floral (RF) flavor (*8*), nutty flavor (*4*), and mothball/grassy (MG) flavor. Each example provides a distinct demonstration of linking instrumental and sensory analyses.

Methods

Cheeses

Blocks (18 kg) of commercial Cheddar cheese (4-36 months old, moisture content: 36.0-38.0 %, pH 5.3-5.9) were collected from domestic and international sources and screened for rosy/floral (RF), nutty, or mothball/grassy (MG) flavors by a descriptive sensory analysis panel (described below). The age range of cheeses was specific to the flavor: > 9 months for RF and nutty flavors, and 4 to 24 months for MG flavor based on previous sensory research (*9, 10*).

Chemicals

Diethyl ether (anhydrous, 99.8 %), sodium chloride (99 %), sodium sulfate (99 %) and 2-methyl-3-heptanone (internal standard for quantitation of neutral/basic volatile compounds) were purchased from Aldrich Chemical Company (St. Louis, MO) and 2-methylpentanoic acid (internal standard for quantitation of acidic volatile compounds) was obtained from Lancaster (Windham, NH). Aroma compounds listed in tables were provided by commercial sources (Aldrich Chemical Co., Sigma, St. Louis, MO, Lancaster, Firmenich Inc., Plainsboro, NJ). Sodium bicarbonate (99.7 %), hydrochloric acid (36.5 %), and acetic acid (no 42) were obtained from Fisher Scientific (Pittsburgh, PA).

Descriptive sensory analysis

A trained sensory panel evaluated the selected cheeses using a lexicon developed for Cheddar cheese flavor (*9*). Definitions and references for the terms used are given in Table I. The panelists each received a minimum of 75 h training on descriptive analysis of cheese flavor using the Spectrum™ descriptive analysis method (*11*). Panelists evaluated cheeses individually in booths with free access to bottled water and unsalted crackers for palate cleansing. Panelists expectorated cheeses following analysis. Cheeses were presented in 2 x 2 cm cubes with three digit codes at 10°C. Cheeses were evaluated in triplicate by each panelist.

Table I. Cheddar Cheese Lexicon and References

Term	*Definition*	*References*
Cooked/milky	Aromatics associated with cooked milk	skim milk heated to 85°C for 30 min
Whey	aromatics associated with Cheddar cheese whey	fresh Cheddar whey
Diacetyl	Aromatics associated with diacetyl	diacetyl (2,3-butanedione)
Milkfat/lactone	Aromatics associates with milkfat	fresh coconut meat, heavy cream, δ-dodecalactone
Sulfur	aromatics associated with sulfurous compounds	boiled mashed egg, struck match
Brothy	aromatics associated with boiled meat or vegetable stock	Knorr beef broth cubes, Wyler's low-sodium beef broth cubes, canned potatoes, 2-methyl-3-furanthiol
Free fatty acid	aromatics associated with short chain fatty acids	butanoic acid
Fruity	Aromatics associated with different fruits	fresh pineapple, canned pineapple juice
Nutty	the nut-like aromatic associated with different nuts	lightly toasted unsalted nuts, wheat germ, unsalted Wheat Thins
Catty	Aromatics associated with tom cat urine	2-mercapto-2 methyl-pentan-4-one, 20 ppm
Cowy/phenolic	aromatics associated with barns and stock trailers	*p*-cresol, Band-aids, phenol
Fecal/mothball	Aromatics associated with feces and mothballs	skatole, indole, naphthalene

Rosey/floral	aromatics associated with flowers, generally rose-like, detected in aftertaste	
Sweet	fundamental taste sensation elicited by sugars	sucrose (5% in water)
Salty	fundamental taste sensation elicited by salts	sodium chloride (0.5% in water)
Sour	fundamental taste sensation elicited by acids	citric acid (0.08% in water)
Bitter	fundamental taste sensation elicited by caffeine, quinine	caffeine (0.08% in water)
Umami	fundamental meaty taste elicited by monosodium glutamate (msg)	monosodium glutamate (1% in water)

Adapted from reference *9*.

Extraction of Volatile Compounds

For all three flavor studies, selected cheeses were first extracted by direct solvent extraction (DSE) with diethyl ether after addition of internal standards (20 uL of a solution prepared by the addition of 50 μL of 2-methyl-3-heptanone and 50 μL of 2-methyl pentanoic acid to 5 mL of methanol) followed by high vacuum distillation (HVD) (nutty, MG flavor studies) (or solvent assisted flavor evaporation (SAFE) (rosy/floral flavor study). Details for HVD and SAFE are provided elsewhere (*4, 8*). Concentrated distillates were fractionated into neutral/basic (NB) and acidic (Ac) fractions for GC separation and identification (*4, 8*).

Gas Chromatography-Mass Spectrometry (GC-MS)

GC-MS analysis of solvent fractions was performed using an HP5890 Series II GC/HP 5972 mass selective detector (MSD, Hewlett-Packard, Co.) system. Volatile compound separations were performed on a fused silica capillary column (DB-WAX; 30 m length x 0.25 mm i.d. x 0.25 μm d_f, J&W Scientific). Helium gas was used as a carrier at a constant flow of 1 mL/min. Oven temperature was programmed from 40°C to 200°C at a rate of 5°C/min with initial and final hold times of 5 and 45 min, respectively. MSD conditions were as follows: capillary direct interface temperature, 280°C; ionization energy, 70 eV; mass range, 33-330 a.m.u; EM voltage (Atune+200 V); scan rate, 5 scans/s. Each solvent fraction (2 μL) was injected in the splitless mode (*8*). Duplicate analyses were performed on each sample.

Gas Chromatography/Olfactometry (GC-O)

Two different GC-O techniques were utilized to evaluate aroma properties and perceived sensory intensities of each aroma-active component in the solvent fractions: post peak intensity (*12*) and aroma extract dilution analysis (AEDA) (*12, 13*). GC-O was performed using a HP5890 series II gas chromatograph (Hewlett-Packard Co., Palo Alto, CA) equipped with a flame ionization detector (FID), a sniffing port, and splitless injector was used. Both the neutral/basic and acidic fractions were analyzed from every solvent extraction. Each fraction (2 μL) was injected into a polar capillary column (DB-WAX; 30 m length x 0.25 mm i.d. x 0.25 μm film thickness d_f; J & W Scientific, Folson, CA) and a nonpolar column (DB-5MS; 30 m length x 0.25 mm i.d. x 0.25 μm d_f; J & W Scientific, Folson, CA). The column effluent was split 1:1 between the FID and the sniffing port using deactivated fused silica capillaries (1 m length x 0.25 mm i.d.; Restek, Bellefonte, Pa., U.S.A.). The GC oven temperature was programmed from 40 °C to 200 °C at a rate of 10 °C/min with an initial hold for 3 min and a final hold of 20 min. The FID and sniffing port were maintained at a temperature of 250 °C. The sniffing port was supplied with humidified air at 30 mL/min flow rate. The fractions were diluted stepwise with diethyl ether at a ratio of 1:3 (v/v). Four experienced sniffers, each with greater than 50 h training on GC/O, were used for post peak intensity GC-O, and two were used for AEDA.

Compound identification

For identifications, retention indices (RI), mass spectra, and odor properties of unknown compounds were compared with those of authentic standard compounds analyzed under identical conditions. Tentative identifications were based on comparing mass spectra of unknown compounds with those in the National Institute of Standards and Technology (1992) mass spectral database or on matching the RI values and odor properties of unknown compounds against those of authentic standards. For the calculation of retention indices, an n-alkane series was used (*14*).

Dynamic Headspace Analysis (DHS)

For two of the cheese flavors, nutty and mothball/grassy (MG) flavors, DHS was conducted as a secondary volatile compound extraction approach. For the dynamic headspace sampling, grated cheese sample (10 g), in duplicate, was first equilibrated in a 3-neck glass purge vessel (280 mL volume, jacketed; Custom Glass Shop) to 45 °C for 20 min followed by purging of headspace volatiles on to adsorbent Tenax TA 60/80 (200 mg/trap) using nitrogen (flow rate 50 mL/min, ultra high purity). GC-O and dynamic headspace dilution analysis (DHDA) was performed by varying headspace purge times (25, 5, 1 min), as described by Cadwallader and Baek (*15*). Volatiles adsorbed on the Tenax were thermally desorbed in splitless-mode (Thermal Desorption system TDS2, Gerstel GmbH & Co. KG, Germany), and cryo-focused before injection (-150 °C, solvent venting mode; cooled injection system CIS4, Gerstel GmbH & Co.) for GCO analysis. The GC-O system consisted of an HP6890 series GC (Agilent Technologies Inc.) equipped with a DB-FFAP capillary column (15 m x 0.53 mm i.d. x 1 μm film thickness; J&W Scientific, Folsom, CA), an FID, and a sniffing port. The oven temperature was programmed from 30 to 225 °C at a heating rate of 10 °C/min with initial and final hold times of 2 and 10 min, respectively. Helium was used as a carrier gas at 10 mL/min flow rate (flow velocity of 70 cm/s). Two experienced sniffers evaluated each sample/purge time combination. A flavor dilution (FD) factor was calculated for each odorant by dividing the highest purge time tested (25 min) by the purge time at which it was last detected by GCO-DHS (e.g. either 25, 5 or 1 min)(*15*). The GC-MS system consisted of an HP6890 series GC/5973 mass selective detector (Agilent Technologies Inc.) and separation of desorbed volatiles was performed on a DB-FFAP capillary column (30 m x 0.25 mm i.d. x 0.25 μm film thickness; J&W Scientific, Folsom, CA). Oven temperature was programmed from 20 to 225 °C at 4 °C/min with initial and final hold times of 5 and 15 min, respectively. Helium was used as a carrier gas at 1.2 mL/min. MSD conditions were as follows: capillary direct interface temperature, 280 °C; ionization energy, 70 eV; mass range, 35-300 amu; scan rate, 5.27 scans/s.

Sensory threshold and cheese model system analysis

Best estimate orthonasal thresholds (BETs) of selected compounds were determined using the ASTM ascending forced choice method of limits procedure E679-79 (*16*). Thresholds of compounds were determined orthonasally in deodorized water and/or pH 5.5 buffer. The buffer was prepared using 250 mL of 0.2M tris acid maleate (24.2g tris and 23.2g maleic acid/L) and 66 mL of 0.2N NaOH made up to one L in deodorized water. Deodorized water was prepared by taking deionized water and boiling to 2/3 of the original volume. Thresholds can vary widely from person to person and within one individual (*11*). Thus, for an accurate estimation of a threshold, more than 30 individuals should be tested. In these studies, a minimum of 50 individuals were assessed.

Sensory analysis of cheese models was conducted to pinpoint the compound(s) responsible for the specific flavor attributes of selected Cheddar cheeses. Cheese models were prepared from commercial Cheddar cheeses which did not have the specific target flavor, purchased from a local grocery store. Chemicals tested for their flavor impact in the models were dissolved in methanol (aroma evaluation) or 95 % ethanol (flavor evaluation) across the concentration range found in the cheeses. The cheeses were grated and portioned (25 g) and the chemicals were introduced by a clean, disposable micropipet. After addition of the chemicals, the cheese models were kneaded for 3 min and then molded to a rectangular shape, wrapped in foil and equilibrated for 24 h at 5 °C. Cheese models were evaluated for aroma or flavor by sensory analysis using the same procedure applied for descriptive analysis of Cheddar cheeses.

Results and Discussion

Linking sensory and instrumental analysis

Case 1 – Rosy/floral (RF) flavor

Results from this study were straight-forward and demonstrate how odor-active compounds responsible for a particular flavor may be readily selected based on the complete understanding of the sensory perception of the flavor (through the use of a defined and anchored sensory language) and close examination of the GC-O results. RF flavor is described by trained panelists as a distinct rosy/floral aftertaste following consumption (or expectoration) of certain aged Cheddar cheeses (*9*). Examination of instrumental results from aged Cheddar cheeses with and without RF flavor revealed 50 aroma-active compounds (*8*). Several compounds had relatively high $\log_3$ FD (flavor dilution) values in cheeses with and without RF flavor, including butanoic acid (cheesy/rancid), methional (potato), dimethyl trisulfide (cabbage), phenylacetaldehyde (rosey), 2-phenethanol (rosey), 2-methoxyphenol (smoky), 2-acetyl-2-thiazoline (popcorn), and phenyl ethyl acetate (rosey) (Table II). FD factors are generated from aroma extract dilution analysis (AEDA). Fractions are serially diluted stepwise (usually 1:2 or 1:3) and sniffed. The dilution procedure is repeated until sniffers do not detect odorants. The highest dilution at which an odorant is perceived is reported as the $\log_3$ flavor dilution (FD) factor (12). As such, AEDA results (list of FD factors) provide a snapshot or profile of the most potent aroma (flavor) contributing compounds in a given extracted sample. The higher the FD factor, the larger the role that the compound plays in flavor. AEDA is a semi-quantitative technique and therefore the values do not represent actual concentrations of compounds, only their aroma activity in the extracts.

The compounds phenylacetaldehyde, 2-phenethanol, and phenyl acetic acid all have rosy aromas and have been previously associated with unclean flavors in cheeses (*17*). These compounds are formed by the Strecker degradation of aromatic amino acids, especially phenylalanine (*18*). The three compounds were quantified in cheeses with and without RF flavor using external standard curves, and sensory threshold analysis was then conducted. Threshold analysis is often useful in flavor chemistry studies to further screen possible odorants responsible for selected flavors. Compounds present at concentrations well above sensory thresholds are more likely to be key players in flavor contributions (*7*). In RF cheeses, phenylacetaldehyde, 2-phenethanol, and phenyl acetic acid were all above their sensory thresholds (Table III). Phenylacetaldehyde was also above sensory threshold in the non-rosy cheeses, while 2-phenethanol and phenyl acetic acid were near or below threshold concentration levels in the non-rosy cheeses.

Sensory analysis of cheese model systems was conducted to confirm the role of each of the three final candidate compounds: phenylacetaldehyde, 2-phenethanol, and phenyl acetic acid. Alone, each of the three compounds displayed a rosy aroma. However, when spiked into cheese, this observation was not necessarily true. 2-Phenethanol produced a yeasty aroma and flavor to the cheese models when added at the levels found naturally in RF Cheddar cheese (200 to 600 ppb) (*8*). Phenylacetaldehyde and phenyl acetic acid contributed to RF flavors in Cheddar cheese (Figure 1). Though phenylacetaldehyde (9.5/10 overall similarity by sensory analysis compared to 7.8/10 overall similarity for phenyl acetic acid) alone exhibited a close similarity to the RF flavor found naturally in cheese, when phenyl acetic acid and phenylacetaldehyde were combined, they contributed more intensity of this flavor to the cheese. Even at the lowest levels found in RF cheese, the combination of these two compounds play a role in RF flavor. This study provides an excellent example of a straight-forward relationship between

Table II. Aroma-active Components in Cheeses With and Without Rosy/Floral (RF) Flavor

no.	*compound*	*fraction*	*odor* [a]	*RI* [b]		*Log* $_3$ *FD Factor* [c]			
				DB-5	*WAX*	*r-1* [d]	*r-2* [d]	*nr-1* [d]	*nr-2* [d]
1	acetic acid [f]	Ac	vinegar	600	1393	2	5	5	3
2	2,3-butanedione (diacetyl) [f]	NB	buttery	670	956	5	3	4	4
3	3-methyl butanal [f]	NB	malty	686	925	1	5	3	ND [i]
4	2-methyl butanal [f]	NB	dark chocolate	688	ND	<1	<1	3	<1
5	3-hydroxy-2-butanone [f]	NB	buttery	715	990	1	1	3	2
6	ethyl butyrate [g]	NB	bubblegum	730	1000	ND	<1	4	4
7	hexanal [f]	NB	green/sweet	810	1048	3	4	<1	<1
8	unknown	NB	skunk	827	1067	<1	6	3	5
9	butanoic acid [f]	Ac	rancid cheese	ND	1610	5	7	7	6
10	(*Z*)-3-hexenal [g]	NB	fruity/solvent	857	1037	2	4	2	2
11	2-methyl-3-furanthiol [g, h]	NB	brothy/vitamin	875	ND	2	4	3	2
12	(*Z*)-4-heptenal [g]	NB	fatty/fishy	903	1220	<1	2	2	<1
13	pentanoic acid [f]	Ac	sour/cheesy/beefy	925	1713	3	ND	<1	ND
14	methional [f]	NB	potato	911	1429	4	8	7	7
15	2-acetyl-1-pyrroline [g]	NB	popcorn	925	1317	<1	4	<1	<1
16	methyl hexanoate [g]	NB	citrus	938	1118	<1	<1	1	<1
17	unknown	NB	burnt	962	ND	<1	2	ND	<1
18	dimethyl trisulfide [g]	NB	cabbage/sulfur	973	1362	2	6	1	<1
19	1-octen-3-one [g]	NB	metallic/mushroom	983	1270	2	5	3	3
20	ethyl hexanoate [f]	NB	fruity/citrus	996	1221	3	6	3	2
21	octanal [f]	NB	green/citrus	1008	1273	1	<1	1	ND
22	hexanoic acid [f]	Ac	sweaty	1019	1861	<1	<1	2	<1
23	phenylacetaldehyde [f]	NB	floral/honey	1044	1619	3	6	3	3
24	2,5-dimethyl-4-hydroxy-3(2*H*)-furanone (Furaneol) [g]	Ac	burnt sugar	1071	2054	3	3	3	<1
25	2-methoxyphenol (guaiacol) [f]	NB	burnt/smoky	1092	1480	1	7	3	7

26	nonanal [f]	NB	fatty/floral	1100		4	3	3	3
27	2-acetyl-2-thiazoline [g]	NB	popcorn	1111	1793	5	7	3	3
28	3-hydroxy-4,5-dimethyl-2(*5H*)-furanone (sotolon) [g]	Ac	curry/maple/spicy	1127	2210	5	5	4	<1
29	(*Z*)-2-nonenal[g]	NB	green/floral	1130	1528	2	5	ND	ND
30	2-ethyl-4-hydroxy-5-methyl-3(*2H*)-furanone (homofuraneol) [g]	Ac	burnt sugar	1142	ND	5	4	3	2
31	2-phenethanol [f]	NB	floral/rosey	1150	ND	9	8	4	6
32	(*E, Z*)-2,6-nonadienal [g]	NB	cucumber	1154	1573	5	2	1	<1
33	(*E*)-2-nonenal [f]	NB	old books/paper	1164	1568	5	6	3	4
34	(*E,E*)-2,4-nonadienal [g]	NB	fatty	1224	ND	1	4	2	<1
35	phenyl ethyl acetate [g]	NB	fatty/rosey	1260	ND	8	7	3	6
36	phenyl acetic acid [f]	Ac	honey	1265	2569	<1	<1	<1	<1
37	δ-octalactone [f]	NB	coconut	1287	ND	1	4	1	4
38	(*E,Z*)-2,4-decadienal [g]	NB	fatty	1316	ND	ND	ND	3	ND
39	o-aminoacetophenone [g]	NB	grape/stale	1315	ND	3	2	2	4
40	(*E,E*)-2,4-decadienal [g]	NB	fatty	1325	ND	2	4	3	3
41	γ-nonalactone [g]	NB	coconut/cilantro	1360	ND	<1	3	3	ND
42	unknown	NB	grain/floral	1397	1745	3	4	2	3
43	3-methyl indole [g]	NB	fecal/mothball	1399	ND	2	5	5	ND
44	Vanillin[g]	Ac	vanilla	1430	ND	ND	<1	<1	6
45	δ-decalactone[g]	NB	peach	1481	ND	5	<1	3	2
46	γ-decalactone[g]	NB	coconut	1508	ND	<1	5	5	4
47	6-(*Z*)-dodecen γ-lactone[g]	NB	soapy/waxy	1603	ND	6	5	5	4
48	δ-dodecalactone[g]	NB	peach	1705	ND	5	4	5	6
49	nonanoic acid[g]	Ac	sweaty	ND	2053	1	3	ND	<1
50	unknown	Ac	caramelized/maple	ND	>2400	4	3	<1	1

[a] Ac- acidic fraction, NB – neutral/basic fraction. [b] Odor description at GC-sniffing port. [c] Retention indices (RI) from GC-O data. [d] Flavor dilution factors determined on DB-5MS column for NB compounds, and on DB-WAX column for Ac compounds. [e] r = rosey, nr = non-rosey. [f] Compound positively identified by comparison of RI and MS data and odor characteristics with the authentic standard. [g] Compound tentatively identified by comparison of RI data and odor characteristics with authentic standard. [h] Compound not previously identified in Cheddar cheese. [i] Compound not detected. (Adapted from reference *8*)

Table III. Concentrations and Sensory Orthonasal Threshold Values of Rosy/Floral (RF) Compounds

Compound	*RI*	*concentration (ppb)*				*experimental threshold (ppb)*	
		RF 1	*RF 2*	*Non RF 1*	*Non RF 2*	*water*	*pH 5.5 buffer*
phenylacetaldehyde	1044[a]	3700 ± 1200	10000 ± 3200	1300 ± 200	1900 ± 900	2 ± 0.8	0.8 ± 0.7
2-phenethanol	1150[a]	600 ± 300	200 ± 10	100 ± 100	400 ± 300	122 ± 0.9	176 ± 0.9
phenyl acetic acid	1265[b]	400 ± 300	40 ± 30	1 ± 2	30 ± 40	464 ± 0.9	17 ± 0.8

[a] Retention index (RI) calculated from mass spectrometry (MS) results on a DB-5MS column. [b] Retention index calculated from FID results on a DB-WAX column. Adapted from reference *8*.

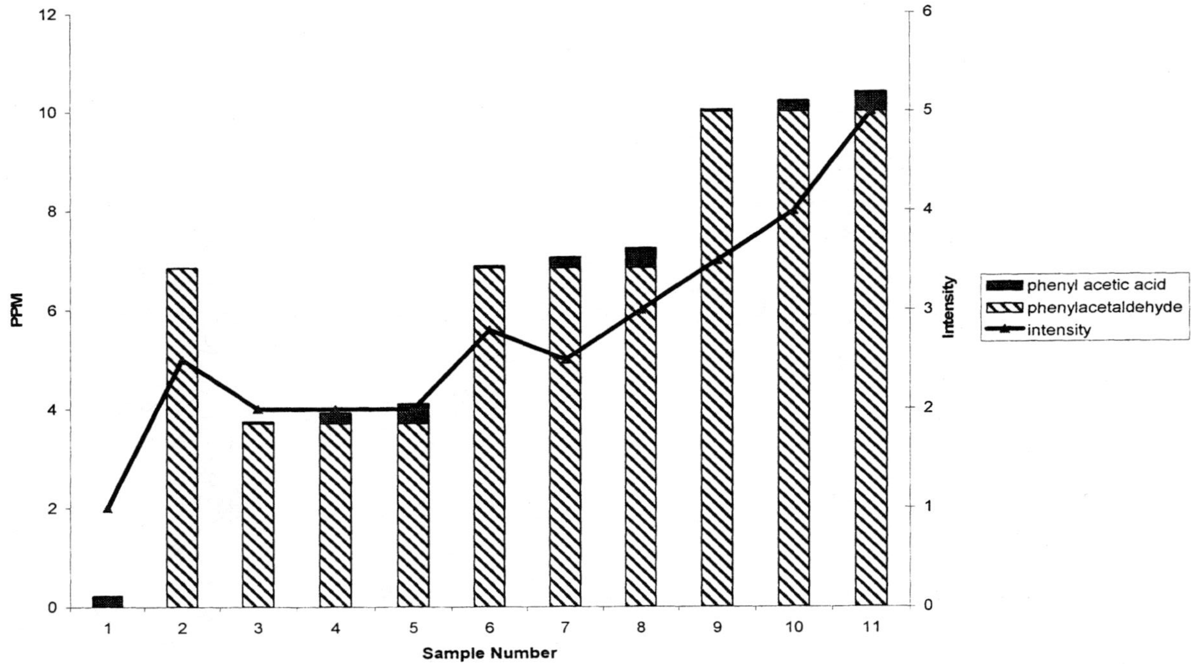

Figure 1. Impact of the addition of phenylacetaldehyde and phenyl acetic acid on rosy/floral (RF) flavor of cheese models. [Intensity refers perceived sensory intensity of RF flavor.] (Reproduced from reference 6. Copyright 2005 American Chemical Society.)

compound(s) and the sensory perception of a flavor. In this case, the compounds that contributed to RF flavor were readily recovered by the instrumental extraction approach, and GC-O aroma of the individual compounds was characteristic of the flavor they produced in Cheddar cheese. Other cheese volatile components linked to specific flavors (waxy/crayon, cowy/barny) also displayed similar relationships (*8, 19*).

Case 2 - nutty flavor

The search for nutty flavor sources in Cheddar cheese provides us with a more difficult case. In this study, cheese with high intensities of nutty flavor were carefully selected along with cheeses of similar age without these flavors. This flavor only occurs at high intensities in Cheddar cheeses that are aged (>9 mo). However after solvent extraction and characterization of volatile components, no clear differences in aroma-active components were observed between nutty and not-nutty cheeses (*4*). Table IV shows the aroma-active compounds found in both nutty and not-nutty cheeses from solvent extraction with high vacuum transfer. Relative abundance calculations (data not shown) also revealed no consistent differences between nutty and not-nutty cheeses.

This study is an excellent example of the shortcomings of a single extraction approach for volatile compounds. No single extraction approach successfully isolates all of the volatile components from a food matrix (*16*). Solvent extraction approaches are excellent for isolation of medium and higher molecular weight volatile components, but are less effective at low molecular weight (highly volatile) compounds (*2*). In the case of nutty flavor, application of an alternative extraction approach proved useful. Dynamic headspace analysis (DHA) revealed consistent qualitative and quantitative differences between nutty and not-nutty cheeses in 2-methyl propanal, 2-/3-methyl butanal and dimethyl sulfide (Table V) (*4*). Higher relative abundances of 2-methylpropanal and 2-/3-methylbutanal were also noted in nutty Cheddar cheeses compared to not nutty Cheddar cheeses by quantitative instrumental analysis (*4*). Strecker aldehydes, particularly 2-methyl propanal, were generally present between 500 to 1000 ppb in aged Cheddar cheeses with nutty flavors and less than 500 ppb in cheeses without nutty flavor.

Sensory analysis of cheese model systems was used to confirm the role of 2-methyl propanal and 2-/3-methyl butanal in nutty flavor. Each Strecker aldehyde was spiked into mild or aged not nutty Cheddar cheese across the concentration range found in naturally nutty Cheddar cheeses (500 to 1000 ppb). Descriptive sensory analysis of these cheeses revealed that in mild Cheddar cheeses, Strecker aldehydes caused malty flavors, a result consistent with the actual aroma of these pure compounds (*4*). However, when placed into aged Cheddar cheeses (>8 mo) that already exhibited typical aged flavors such as sulfur and brothy notes, Strecker aldehyde addition resulted in nutty flavors. In this case, the actual aroma of the pure compound was not identical to the flavor the compound elicited in the food matrix. The presence of other compounds at or below sensory threshold and the complex matrix composition (e.g. fat, moisture, pH, ionic strength) can play a large role in the final perception of a particular compound in a food matrix (*7*). This underlies the need to conduct all three steps (sensory analysis, instrumental analysis, sensory analysis of model systems) in linking volatile compounds to sensory perception of a flavor in a food. GC-O results alone may sometimes be misleading.

Case 3 – Mothball/grassy (MG) flavor

MG flavor, also called fecal/mothball flavor, is a flavor that has been noted in international Cheddar cheeses and Cheddar cheeses made from grass fed cows (*10*). The flavor is present in both young and aged cheeses and can be readily detected orthonasally in the headspace when cheese is placed in a sample cup as well as when the cheese is placed in the mouth. Two complementary extraction procedures, direct solvent extraction with high vacuum distillation (DSE/HVD) and dynamic headspace analysis in conjunction with GC-O and GC-MS were applied to young and aged Cheddar cheeses with and without MG flavor. Selected compounds were quantified using external standard curvesby spiking the compounds into deodorized water followed by DSE/HVD and GC-MS analysis.

Table VI shows the sensory profiles of the cheeses with and without MG flavor. MG flavor does not appear to depend on cheese age. There were no differences between MG and non MG cheeses as determined by dynamic headspace (data not shown). Many compounds (57) (Table VII) were identified with Log_3 FD factors of >1 in the cheeses. Log_3 FD factors ranged from 1-9 in these cheeses. Some compounds which were detected with high FD factors in all the cheeses included acetic acid (vinegar), butanoic acid (cheesey) 1-nonen-3-one (metallic/beefy), 2-phenethanol (rosy), and γ-octalactone (coconut).

3-Methyl indole had a higher Log_3 FD factor by AEDA in cheeses with MG flavor than in the control cheeses. This compound is commonly found in the milk from pasture fed animals (*20*) and is formed from aromatic amino acid catabolism, specifically from the degradation of tryptophan (*21*). The presence of 3-methyl indole is common in aged Cheddar cheese (*22, 23*). 3-Methyl indole has a mothball aroma. Sensory panelists indicated that the aroma of 3-methyl indole was not identical to MG flavor in Cheddar cheese, and therefore more compounds may be contributing to MG flavor in Cheddar cheese.

Seven compounds were selected for quantitification based on high FD factors and aroma properties (Table VIII). Acetic acid, 2-methyl butanoic acid, and 3-methyl indole all had higher concentrations and Log_3 FD factors in cheeses with MG flavor compared to control cheeses. Sensory analysis of cheese model systems was conducted to determine if any or all three of these compounds contributed to MG flavor. Compounds (acetic acid, 2-methyl butanoic acid, 3-methyl indole) were spiked into cheeses without MG flavor.

Table IV. Aroma-active Components in Nutty and Not Nutty Cheddar Cheeses

no.	Compound	fraction[a]	odor[b]	RI[c]		Odor intensity (5 point intensity scale)			
				DB-5	WAX	NN1[d]	NN2[d]	N1[d]	N2[d]
1	acetic acid[d]	Ac	vinegar	600	1393	3.9	--	3.0	4.2
2	2,3-butanedione (diacetyl)[d]	NB	buttery	670	956	1.3	0.5	0.25	3.5
3	1-hexen-3-one[e]	NB	rubbery	747	1099	1	--	1.5	--
4	ethyl butyrate[d]	NB	bubblegum	730	1000	2.8	2.3	1.8	2.0
5	hexanal[d]	NB	green/sweet	810	1048	1.8	--	--	0.5
6	unknown	NB	Plastic, coffee	829		--	1.0	2.0	1.5
7	butanoic acid[d]	Ac	rancid cheese	844	1620	--	2.3	3.3	3.5
8	2-methyl-3-furanthiol[e]	NB	brothy/vitamin	875		1.5	4.0	--	3.8
9	unknown	NB	Soil, mushroom	890		--	--	1.5	--
10	(*Z*)-4-heptenal[e]	NB	fatty/fishy	903	1220	--	1.0	--	--
11	propionic acid[d]	AC	sour	916	1458	--	--	3.5	4.0
12	pentanoic acid[d]	Ac	sour/cheesy/sweaty	925	1713	5.8	--	3.0	4.0
13	methional[d]	NB	potato	911	1450	5.0	3.8	3.3	8.0
14	2-acetyl-1-pyrroline[d]	NB	popcorn	925	1317	0.5	--	0.75	5.0
15	ethyl pentanoate[e]	NB	fruity	929	1137	--	2.3	--	--
16	isobutyric acid[d]	AC	Sour	966	1520	6.4	2.9	2.0	4.2
17	dimethyl trisulfid[e]	NB	cabbage/sulfur	973	1362	--	1.0	--	--
18	1-octen-3-one[e]	NB	metallic/mushroom	983	1270	3.5	2.8	2.0	4.5
19	ethyl hexanoate[d]	NB	fruity/citrus	996	1221	0.5	1.0	2.3	1.5
20	unknown	NB	Nutty, earthy	1005		0.8	--	2.3	4.5
21	(*E*)-2-octenal[e]	NB	Dirty/stale	1039		--	--	0.5	1.5
22	hexanoic acid[d]	Ac	sweaty	1035	1861	2.5	2.0	6.0	5.9
23	phenylacetaldehyde[d]	NB	floral/honey	1044	1619	2.0	--	--	4.5
24	maltol[d]	AC	Burnt sugar	1060	1991	--	7.0	--	--

25	tetramethylpyrazine [d]	NB	nutty	1075		--	3.3	2.5	4.5
26	p-cresol [d]	NB	Fecal, bandaid	1084	2078	--	2.0	--	4.5
27	2-isopropyl-3-methoxy-pyrazine [e]	NB	Earthy, nutty	1091	1425	3.5	0.5	5.5	4.5
28	unknown	NB	fatty	1100		3.0	2.5	--	--
29	2-acetyl-2-thiazoline [e]	NB	Popcorn, roasted	1115	1725	3.3	3.0	3.3	5.0
30	sotolon [e]	AC	curry	1145	2196	3.4	3.2	4.5	6.2
31	2,3-diethyl-5-methylpyrazine [e]	NB	Sweet, nutty	1151		--	0.8	0.8	3.5
32	(*E,Z*)-2,6-nonadienal [e]	NB	cucumber	1153	1583	1.5	1.8	0.5	2.0
33	(*E*) 2-nonenal [e]	NB	Hay, tobacco	1168	1535	1.0	2.8	3.0	2.5
34	2-isobutyl-3-ethoxypyrazine [e]	NB	nutty	1192		1.0	2.0	4.5	4.5
35	δ-octalactone [d]	NB	coconut	1306	1977	1.3	2.3	2.0	3.3
36	(*E,E*)-2,4-decadienal [e]	NB	fatty	1332	1313	2.5	--	--	--
37	γ-nonalactone [d]	NB	coconut/cilantro	1360	2020	1.0	2.0	2.0	1.0
38	3-methyl indole [d]	NB	fecal/mothball	1473	2489	--	1.3	--	5.0
39	δ-decalactone [d]	NB	Peach, coconut	1481	2179	0.5	--	2.8	2.5
40	γ-decalactone [d]	NB	coconut	1508	2366	--	--	0.5	4.0
41	6-(*Z*)-dodecen γ-lactone [d]	NB	soapy/waxy	1650		--	--	2.5	2.5
42	δ-dodecalactone [d]	NB	peach	1705		2.0	0.5	1.0	1.3

[a] Ac- acidic fraction, NB – neutral/basic fraction. [b] Odor description at the GC-sniffing port. [c] Retention indices calculated using an n-alkane series. [d] N = nutty, NN = not nutty. [e] Compound identified by comparison of RI and MS data and odor characteristics with the authentic standard. [f] Compound tentatively identified by comparison of RI data and odor characteristics with authentic standard. (Adapted from reference *4.*)

Table V. Aroma-active Components of Not-Nutty (NN) and Nutty (N) Cheddar Cheeses by Gas Chromatography-Olfactometry of Dynamic Headspace Dilution Analysis

Compound	*RI*[a] *(FFAP)*	*Odor*[b]	*FD-factor*[c] *(odor intensity) 5 point intensity scale*		
			NN3	*NN4*	*N3*
dimethylsulfide	706	Cut cabbage	5 *(1)*	5 *(1)*	25 *(3)*
2-methyl propanal	770	Dark chocolate	-	5 *(3)*	25 *(3)*
2-/3-methy lbutanal	915	Dark chocolate	25 *(1)*	25 *(1)*	25 *(4)*
2,3-butanedione	980	Buttery	25 *(1)*	25 *(1)*	25 *(3)*
ethyl butyrate	1052	Fruity, bubble gum	25 (3)	25 (1)	25 (3)
ethyl hexanoate	1244	Fruity, berry-like	1 (2)	5 (1)	25 (3)
1-octen-3-one	1320	Mushroom	25 (2)	25 (1)	25 (3)
dimethyltrisulfide	1386	Sulfurous, cabbage	25 (2)	25 (1)	25 (4)
methional	1497	Potato	25 (2)	25 (2)	25 (4)

[a] Retention indices (RI) on DB-FFAP calculated from GCO results. [b] Odor description at the GC-sniffing port during GC. [c] Flavor dilution factor =highest purge time tested (25 min) divided by lowest purge time in which odorant was last detected by GCO (either 25, 5, or 1 min). Numbers in parentheses represent post peak odor intensities. (Reproduced from reference *4*. Copyright American Dairy Science Association.)

Table VI. Sensory Analysis of Cheeses With and Without Mothball/Grassy (MG) Flavor

Attribute	*6 mo control*	*6 mo MG*	*12 mo control*	*12 mo MG*	*15 mo control*	*15 mo MG*
cooked	2.70^{a}	2.68^{a}	3.00^{a}	3.00^{a}	1.5^{b}	1.75^{b}
whey	2.17^{a}	2.36^{a}	ND	ND	ND	ND
diacetyl	1.75^{a}	1.64^{a}	ND	ND	ND	ND
milkfat/lactone	3.78^{a}	3.99^{a}	3.00^{b}	3.00^{b}	3.0^{b}	1.76^{c}
cowy/barny	ND	1.00^{a}	ND	ND	ND	1.08^{a}
Mothball/grassy	ND	2.69^{b}	ND	4.25^{a}	ND	2.47^{b}
sulfur	1.76^{b}	0.76^{c}	2.50^{b}	2.50^{b}	4.00^{a}	2.6^{b}
brothy	2.05^{b}	1.26^{c}	3.00^{ab}	2.50^{ab}	3.65^{a}	2.21^{b}
nutty	0.69^{a}	0.32^{a}	ND	ND	ND	1.01^{a}
catty	ND	ND	ND	ND	2.00	ND
sour	2.86^{b}	2.23^{b}	3.00^{ab}	3.00^{ab}	3.50^{a}	2.95^{b}
bitter	ND	ND	ND	ND	1.50^{a}	0.55^{b}
salty	3.42^{a}	3.55^{a}	4.00^{a}	4.00^{a}	2.90^{a}	3.60^{a}
sweet	1.68^{b}	1.66^{b}	2.50^{a}	2.50^{a}	2.00^{b}	2.66^{a}
umami	1.63^{b}	1.41^{b}	2.00^{a}	2.00^{a}	2.75^{a}	1.90^{a}

Means in a row followed by different letters are different ($p<0.05$). Intensities were scored on a 10-point scale where 0 = none and 10 = very high. ND=not detected

Table VII. Potent Odorants in Cheddar Cheeses With and Without Mothball/Grassy (MG) Flavor

no.	compound	fraction[a]	odor[b]	RI[c] DB-5	RI[c] Wax	Log$_3$ FD Factor[d] 6mo C[e]	6mo MG[e]	12mo C[e]	12mo MG[e]	15mo C[e]	15mo MG[e]
1	acetic acid[f]	Ac	vinegar	600	1393	4	7	3	5	4.5	8
2	2,3-butanedione[f]	NB	buttery	670	956	2.5	3	6	<1	<1	<1
3	unknown	NB	sweet/solvent	700	929	0.5	ND[h]	<1	ND	<1	3
4	ethyl butyrate[g]	NB	bubblegum	730	1000	<1	ND	3	4	<1	4
5	2/3-methylbutanoic acid[g]	Ac	sweaty/sweet	ND	1643	5	4.5	ND	1	ND	7
6	butanoic acid[f]	Ac	rancid cheese	ND	1610	5.5	7	5	8	5.5	8
7	2-methyl-3-furanthiol[g]	NB	urine/nutty	875	1037	<1	<1	2.5	3	5	<1
8	unknown	NB	soil/geranium	896	1187	1	ND	<1	ND	1.5.	ND
9	pentanoic acid[f]	Ac	sour/cheesy/beefy	925	1713	ND	1	1	3	4.5	2
10	methional[g]	NB	potato	911	1429	1.5	4	7	5	1	5
11	methyl hexanoate[g]	NB	citrus	938	1118	ND	ND	<1	1	1.5	<1
12	1-octen-3-one[g]	NB	mushroom	983	1270	2	<1	5	<1	1.5	2
13	octanal[f]	NB	green/citrus	1008	1273	1	3.5	2.5	2.5	1	4
14	hexanoic acid	Ac	sweaty	1019	1861	ND	<1	2	4	<1	<1
15	phenylacetaldehyde[f]	NB	floral/honey	1044	1619	<1	<1	5.5	4	1.5	5
16	*p*-cresol[f]	NB	fecal/bandaid	1084	2078	ND	ND	ND	0.5	ND	2
17	2-isopropyl-3-methoxypyrazine[g]	NB	soil/bell pepper	1091	1425	1	ND	<1	ND	5	ND
18	2-methoxyphenol (guiacol)[g]	NB	beefy/burnt	1092	1480	ND	2	ND	2.5	2.5	<1
19	nonanal[f]	NB	fatty	1100	ND	ND	<1	1	2.5	2.5	ND
20	unknown	NB	metallic/beefy/cowy	1105	1305	4.5	8	7	5	4.5	5
21	2-acetyl-2-thiazoline[g]	NB	popcorn	1111	1793	3	1	4.5	1	<1	4
22	3-hydroxy-4,5-dimethyl-2(*5H*)-furanone(sotolon)[g]	Ac	curry/maple/spicy	1127	2210	<1	3.5	<1	<1	<1	1.5
23	(*Z,Z*)-3,6-nonadienal[g]	NB	metallic/fatty	1109	1444	2	ND	<1	<1	4.5	5
24	(*Z*)-2-nonenal[g]	NB	green/floral	1130	1528	1.5	ND	ND	ND	1.5	<1

25	2-phenethanol [f]	NB	floral/rosy	1150	ND	<1	7.5	9	8	6	5
26	(*E,Z*)-2,6-onadienal [g]	NB	cucumber	1154	1573	<1	ND	ND	<1	ND	3
27	(*E*)-2-nonenal [g]	NB	cucumber	1165	1427	1	2	4	5	5.5	3
28	unknown	NB	rosy	1174	1522	2	3.5	ND	ND	ND	5
29	unknown	NB	fatty/citrus	1212	1549	<1	0.5	5.5	<1	2	<1
30	unknown	NB	popcorn/fatty	1230	ND	1	3	1	ND	2.5	5
31	unknown	NB	hay/minty/floral	1245	ND	0.5	3.5	1	4.5	1.5	6
32	phenyl ethyl acetate [g]	NB	fatty/rosy	1260	1820	1	1	1	ND	4.5	2
33	phenyl acetic acid [g]	Ac	rosy	1262	2569	2	ND	ND	ND	0.5	1.5
34	unknown	NB	tobacco/hay	1278	1831	<1	2.5	7	4	1	<1
35	δ-octalactone [f]	NB	coconut	1287	1977	<1	2.5	5	3.5	<1	4
36	o-aminoacetophenone [g]	NB	grape/stale	1315	2252	<1	2.5	5	4	4.5	4
37	(*E,E*)-2,4-decadienal [g]	NB	fatty	1325	1313	2	3.5	4.5	7	2.5	5
38	2-undecanone [g]	NB	coconut/hay	1312	1543	<1	<1	4	<1	2.5	<1
39	unknown	NB	soap	1328	ND	ND	0.5	ND	ND	2.5	ND
40	unknown	NB	grain	1335	ND	3	2.5	2.5	1	5	2.5
41	γ-nonalactone [g]	NB	coconut	1360	2020	<1	2	4	<1	2	2
42	unknown	NB	grain/paper	1397	1745	<1	ND	5.5	4	5	5
43	3-methyl indole [g]	NB	fecal/mothball	1399	2489	3.5	5.5	6	7	2.5	5
44	δ-decalactone [g]	NB	peach	1481	2179	2	1	4	1	3.5	3
45	γ-decalactone [f]	NB	coconut	1508	2366	5	7	9	8	6	6
46	6-(*Z*)-dodecen-γ-lactone [f]	NB	soapy/waxy	1603	ND	4.5	3.5	5.5	1.5	4.5	2
47	δ-dodecalactone [f]	NB	peach	1705	ND	4.5	1	6	4	3	5
48	nonanoic acid [f]	Ac	sweaty	1762	2053	1.5	2.5	2	<1	<1	5
49	unknown	Ac	strawberry	1902	2239	4.5	<1	2	2.5	ND	4.5
50	decanoic acid [f]	Ac	musty/sour	ND	2255	ND	ND	ND	ND	ND	4.5
51	unknown	Ac	caramelized/maple	ND	>2400	4.5	4	<1	4.5	5.5	8

[a] Ac- acidic fraction, NB – neutral/basic fraction. [b] Odor description at the GC-sniffing port. [c] Retention indices calculated from GC/O data. [d] Flavor dilution factors determined on a DB-5MS column for NB compounds, and on a DB-WAX column for Ac compounds. [e] C= control cheese without MG flavor, MG – cheese with mothball/grassy flavor. [f] Compound identified by comparison of its RI and MS data and odor characteristics with those of authentic standard compounds. [g] Compound tentatively identified by comparison of its RI data and odor characteristics with those of authentic standard compounds. [h] Compound not detected.

Table VIII. Concentration of Selected Compounds in Cheddar Cheeses With and Without Mothball/Grassy (MG) Flavor.

compound	*RI*[a]	*concentration (ng/g)*						*reported threshold (ng/g)*
		6 mo no MG flavor	*6 mo with MG flavor*	*12 mo no MG flavor*	*12 mo with MG flavor*	*15 mo no MG flavor*	*15 mo with MG flavor*	
acetic acid	600	63000 ± 28000	105000 ± 74000	46000 ± 5000	69000 ± 15000	45000 ± 1900	168000 ± 45000	22000 [b]
butanoic acid	880	5000 ± 2000	8000 ± 4000	6000 ± 200	5000 ± 200	7000 ± 100	3000 ± 700	1000 [b]
2/3-methyl butanoic acid	872	20 ±10	200 ± 100	100 ± 20	200 ± 50	10 ± 1	40 ± 10	250 [b]
2-phenethanol	1160	ND	ND	3 ± 0.004	2 ± 0.3	ND	0.6 ± 0.6	240 [c]
4-ethylbenzaldehyde	1174	ND	ND	ND	ND	ND	0.2 ± 0.2	not reported
(*E,E*)-2,4-decadienal	1350	ND	ND	0.6 ± 0.01	3 ± 4	3 ± 4	7 ± 11	0.2 [b]
3-methyl indole	1425	ND	1 ± 1	ND	3 ± 1	ND	6 ± 2	0.7 [b]

[a] Retention indices (RI) calculated from mass spectrometry (MS) results on a DB-5 column. [b] Orthonasal thresholds in water reported by Rychlik et al. (*24*). [c] Threshold in milk reported by Dunn and Lindsay (*17*).

These compounds alone or in combination, did not completely simulate MG flavor (Table IX). Methional and *o*-aminoacetophenone (*o*-AAP) were detected in the cheeses using AEDA/GCO, but were below detection limits of the mass spectrometer. Therefore these compounds were also spiked into model cheeses at concentrations at and above their orthonasal sensory thresholds. A combination of acetic acid, 2-methyl butanoic acid, 3-methyl indole, methional, and *o*-AAP gave the closest match to MG flavor in Cheddar cheese (8.0/10). It was apparent that acetic acid, 3-methyl butanoic acid, 3-methyl indole, methional, and o-aminoacetophenone all contributed to MG flavor. However, the similarity match was not as high as would be desired and we must conclude that there are potentially other compounds that are contributing to this flavor in Cheddar cheese.

Conclusions

Cheddar cheese flavor is very complex, and there are several unique flavors which can be present. When establishing definitive causal links between volatile components and sensory perception of flavors, several different situations can arise. In some cases, one or two compounds clearly cause a particular flavor, rosy/floral flavor. The compounds when isolated have the same aroma as the flavor that they cause in the cheese. In other cases, the aroma the compound displays when isolated may not be identical to the flavor that it causes in the actual cheese, nutty flavor. Finally, some flavors are caused by multiple compounds and pinpointing these compounds and adequately reproducing the flavor in model systems is challenging on multiple levels, mothball/grassy flavor.

Acknowledgments

Funding was provided by Dairy Management, Inc. and the California Dairy Research Foundation. This is manuscript FSR 06-03 of the Department of Food Science, North Carolina State University. The use of trade names in the publication does not imply endorsement by these organizations nor criticisms of ones not mentioned.

Table IX. Composition and Sensory Perception of Cheddar Cheese Model Systems

Compound (s) [a]	Concentration (s) [a]	Mothball/grassy flavor Intensity [b]	Overall Similarity [c]
acetic acid	14 ppm	ND	0f
2-methyl butanoic acid	150 ppb	0.5c	2e
3-methyl indole	0.9 ppb	1.0bc	2e
acetic acid/2-methyl butanoic acid	14ppm/150ppb	1.5b	4d
acetic acid/2-methyl butanoic acid	28ppm/300ppb	1.5b	4d
acetic acid/3-methyl indole	14ppm/0.9ppb	0.5c	0f
acetic acid/3-methyl indole	28ppm/1.8ppb	1.0bc	0f
2-methyl butanoic acid/3-methyl indole	150ppb/0.9ppb	1.0bc	5d
2-methyl butanoic acid/3-methyl indole	300ppb/1.8ppb	0.5c	5d
acetic acid/2-methyl butanoic acid/3-methyl indole	14ppm/150ppb/0.9ppb	0.5c	5d
acetic acid/2-methyl butanoic acid/3-methyl indole	28ppm/300ppb/1.8ppb	0.5c	5d
acetic acid/2-methyl butanoic acid/3-methyl indole/methional	28ppm/300ppb/1.8ppb/20ppb	0.5c	5d

acetic acid/2-methyl butanoic acid/3-methyl indole/o-aminoacetophenone (o-AAP)	28ppm/300ppb/1.8ppb/0.2ppb	0.5c	5d
acetic acid/2-methyl butanoic acid/3-methyl indole/methional/o-AAP	84ppm/0.1ppm/5.4ppb/20ppb/0.2ppb	0.5c	5d
acetic acid/2-methyl butanoic acid/3-methyl indole/methional/o-AAP	84ppm/0.1ppm/7.2ppb/20ppb/0.2ppb	0.75bc	6c
acetic acid/2-methyl butanoic acid/3-methyl indole/methional/o-AAP	84ppm/0.1ppm/10.8ppb/20ppb/0.2ppb	1bc	6.5c
acetic acid/2-methyl butanoic acid/3-methyl indole/methional/o-AAP	84ppm/0.1ppm/12.6ppb/20ppb/0.2ppb	1.5b	7b
acetic acid/2-methyl butanoic acid/3-methyl indole/methional/o-AAP	84ppm/0.1ppm/21.6ppb/20ppb/0.2ppb	1.5b	7b
acetic acid/2-methyl butanoic acid/3-methyl indole/methional/o-AAP	84ppm/0.1ppm/45ppb/20ppb/0.2ppb	3a	8.5a

[a] Compounds were added at listed concentrations to mild Cheddar cheese (purchased from a local grocery store). [b] Mothball/grassy flavor intensity based on the 15-point Spectrum scale (*11*). [c] Based on a 10-point similarity scale 1=different, 10=same, where the reference was Cheddar cheese with MG flavor. [a,b,c,d,e,f – Means in a column followed by different lower case letters are different ($p<0.05$).]

References

1. Drake, M.A. *J. Dairy Sci.* **2004**, *87*, 777-784.
2. Singh, T.; Drake, M.A.; Cadwallader, K.R. *Compr. Rev. Food Sci. Food Safety* **2003**, *2*, 139-162.
3. Broadbent, J.R.; Gummala, S.; Hughes, J.E.; Johnson, M.E.; Rankin, S.A.; Drake, M.A. *Appl. Environ. Micro.* **2004**, *70,* 4814-4820.
4. Avsar, Y.; Karagul-Yuceer, Y.; Drake, M.A.; Singh, T.; Yoon, Y. Cadwallader, K.R. *J. Dairy Sci.* **2004**, *87,* 1999-2010.
5. McGorrin R.J. In *Food Flavors and Chemistry: Advances of the New Millenniu;* Spanier A.M., Shahidi F., Parliment T.H., Mussinan C.J., Ho C.T., Contis E.T., Eds. Royal Society of Chemistry: Cambridge, UK, 2001; p 67-84.
6. Parliment, T.H.; McGorrin, R.J. In *Flavor Chemistry: Industrial and Academic Research;* Risch S.J., Ho C.T., Eds. ACS Symposium Series 756. American Chemical Society: Washington, DC, 2000; p 44-71.
7 Drake, M.A.; Civille, G.V. Flavor lexicons. *Compr. Rev. Food Sci. Food Safety*, **2003**, *2,* 33-40.
8. Carunchia Whetstine, M.E.; Cadwallader, K.R.; Drake, M.A. *J. Agric. Food Chem.* **2005**, *53*, 3126-3132.
9. Drake, M.A.; McIngvale, S.C.; Gerard, P.D.; Cadwallader, K.R.; Civille, G.V. *J. Food Sci.* **2001**, *66,* 1422-1427.
10. Drake, M.A.; Yates, M.D.; Gerard, P.D.; Delahunty, C.M.; Sheehan, E.M.; Turnbull, R.P.; Dodds, T.M. *Int. Dairy J.,* **2005**, *15,* 473-483.
11. Meilgaard, M.M.; Civille, G.V.; Carr, B.T. In *Sensory Evaluation Techniques*, 3nd ed. CRC Press: Boca Raton, FL, 1999; pp 174-176.
12. Van Ruth, S. *Biomolecular Engineering.* **2001**, *17*, 121-128.
13. Grosch, W. *Tr. Food Sci. Technol.* **1993**, *4,* 68-73.
14. Van den Dool, H.; Kratz, P. *J. Chromatogr.* **1963**, *11*, 463-471.
15. Cadwallader, K.R.; Baek, H.H. In Mussinan C, Contis E, Ho CT, Parliment T, Spanier A, Shahidi F, editors. Proceedings of Ninth International Flavor Conference, George Charalambous Memorial Symposium, Elsevier Science BV: Amsterdam, 1998; p 271-279.
16. Lawless, H.; Harono, C.; Hernandez, S. *J. Sensory Stud.* **2000**, *15*, 437-447.
17. Dunn, H.; Linsday, R. *J. Dairy Sci.* **1985**, *68,* 2859-2874.
18. Urbach, G. *Int. Dairy J.* **1995**, *5*, 877-903.
19. Suriyaphan, O.; Drake, M.A.; Chen, X.Q.; Cadwallader, K.R. *J. Agric. Food Chem.* **2001**, *49*, 1382-1387.
20. Bendall, J. G. *J. Agric. Food Chem.* **2001**, *49,* 4825-4832.
21. Yvon, M.; Rijnen, L. *Int. Dairy J.* **2001**, *11,* 185-201.
22. Milo, C.; Reineccius, G. *J. Agric. Food Chem.* **1997**, *45,* 3590-3594.
23. Zehentbauer, G.; Reineccius, G. *Flav. Fragr. J.* **2002**, *17,* 300-305.

24. Rychlik M.; Schieberle P.; Grosch W. *Compilation of thresholds, odor qualitities, and retention indices of key food odorants*. Deutsche Forschungsanstalt fur Lebensmittelchemie and Instiut fur Lebensmittelchemie der Technischen Universitat Munchen: Garching, Germany, 1998.

Chapter 4

Application of Solid-Phase Microextraction Gas Chromatography–Mass Spectrometry for Flavor Analysis of Cheese-Based Products

Raymond T. Marsili

Marsili Consulting Group, Rockford College, 5050 East State Street, Starr Science Building, Room 224, Rockford, IL 61108

This paper describes how solid-phase microextraction (SPME) can be used as a simple, rapid sample preparation technique prior to GC-MS for the study of flavor impact chemicals in cheese and dairy products. Four specific flavor/off-flavor applications are discussed: (1) the quantitative measurement of 2,4,5-trimethyloxazole, a potent off-flavor chemical in spray-dried cultured dairy powder; (2) the quantitation of dimethyl sulfide, the chemical responsible for a desirable creamed-corn flavor in cheese powders; (3) the determination of mold metabolites as the cause of desirable and undesirable flavors in a processed cheese product; and (4) how SPME GC-MS results can be analyzed by chemometrics to predict pass/fail flavor status of cheese powders.

Resolving flavor problems in cheese and other complex dairy products is challenging because of the large number of potential flavor compounds, a lack of understanding of the chemical nature of cheese flavor, the potential for inducing Maillard reaction compounds in sample preparation techniques employed prior to GC-MS analysis and the high fat and protein levels in the dairy/cheese sample matrix. While Solvent Assisted Flavor Evaporation (SAFE), a vacuum distillation-based technique, has been used to extract flavor chemicals from cheese products (*1*), it is too time-consuming and impractical in the context of a busy industrial flavor laboratory which may have dozens of samples submitted each week for flavor profiling.

Dairy chemists have recently exploited the advantages of solid-phase microextraction (SPME) for cheese flavor analysis. Lee et. al (*2*) developed a sensitive and rapid SPME technique using NaH_2PO_4 buffer solution and a poly(dimethylsiloxane)/divinyl benzene SPME fiber.

In the work presented here, Lee's method for SPME GC-MS cheese analysis was modified so it could be applied for use with the Leap Technologies (Carrboro, NC) SPME autosampler, enabling improved precision and greater sample throughput compared to manual SPME methods. Four examples are provided that illustrate how SPME GC-MS can be used to solve significant flavor problems with cheese/dairy products. Also, shown is a chromatogram obtained using SPME GC-TOFMS; the advantages of the application of automatic peak finding and spectral deconvolution algorithms of the Leco TOFMS are illustrated for cheese analysis.

Experimental

In general, the following analytical SPME, GC and MS methods were used:

- Add 1 gm of cheese (cut in small pieces ~1 mm in diameter at -10°C) or cheese powder + 5 mL NaH_2PO_4 (25% w/v) solution to a 20 mL GC vial.
- Vortex for 1 min at high speed.
- Preincubate 4 min at 50°C with agitation.
- Extract 20 min at 50°C with agitation.
- Fiber: poly(dimethylsiloxane)/divinyl benzene
- GC: DB-5 column; 30 m x 0.25 mm; 0.25 μm film thickness.
- GC temperature programming: 40°C for 5 min then heated to 185°C at a rate of 10°C/min and held at 185°C for 5 min.
- CombiPal autosampler with orbital shaker or single magnet stirrer.
- In most cases, GC-MS was performed with an Agilent 6890 GC equipped with a 5973 MSD. The Leco Pegasus III TOFMS was also

employed for some analysis to illustrate the potential benefits of time-of-flight detection for cheese analysis.

Modifications and/or additional steps were made to this general procedure to accommodate the demands of the specific applications.

Off-Flavor in Cultured Sour Cream Powder

Analysis of complaint and control cultured sour cream powder samples consistently showed the presence of 2,4,5-trimethyloxazole (TMO) in complaint samples and significantly less or no detectable TMO in control samples. Sensory testing showed the greater the level of TMO, the more intense the off-flavor. Figure 1 shows a complaint sample of a cultured sour cream dairy powder contaminated with 43 ppb TMO.

Off-flavor problems attributed to TMO have been previously reported in the literature for wine (*3*) and cheese (*4*). Off-flavors attributed to TMO contamination have been described as melon-like, similar to very ripe kiwi notes and extremely pungent. TMO has a taste threshold in water of approximately 5 ppb. Comparison of sensory flavor scores to ppb levels of TMO in sour cream powders showed that the flavor detection of TMO in sour cream powders was approximately 25 ppb.

Formation of TMO has been attributed to the reaction of diacetyl with amino acids (*5*). The reaction of diacetyl with cysteine has been shown to generate TMO, methanethiol and tetramethyl pyrazine in wine samples. Mechanism for formation of TMO from the reaction of diacetyl, acetaldehyde and ammonia has also been proposed. In our laboratory, we have shown that simple mixing of diacetyl with arginine and whey followed by gentle warming is capable of generating significant amounts of TMO. We have also seen TMO-based off-flavors in cheese powders containing high levels of added diacetyl/butter flavors in formulations. In cases where unusual off-flavors occur in dairy products containing high levels of diacetyl and free amino acids, TMO formation should be checked as a possible cause.

Figure 2 shows a typical standard calibration curve for TMO quantitation by the method of addition technique. One-gram samples of sour cream powder with no detectable levels of TMO were spiked with various levels of TMO to generate the calibration curve; no internal standard was employed.

Variability in Creamed-Corn Flavor in Cheese Powders

Variability in a desirable creamed-corn flavor attribute was observed with cheese powders made at different locations. The chemical responsible for the

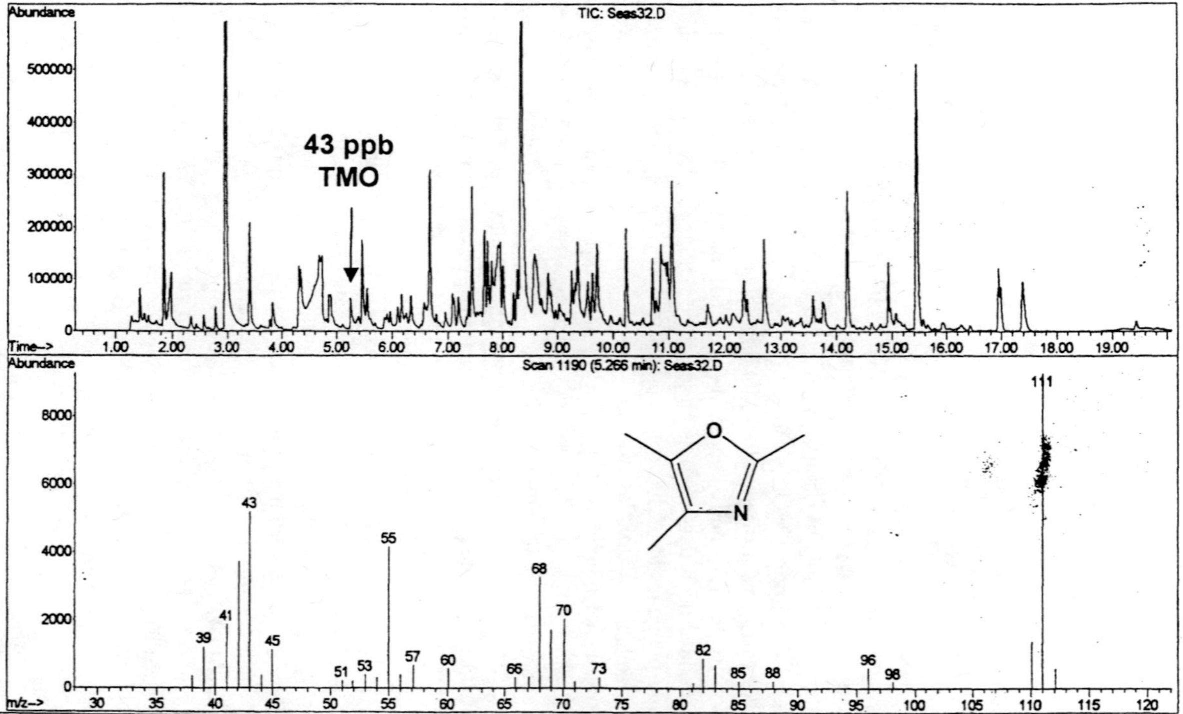

Figure 1. SPME GC-MS (TIC) of sour cream powder showing 43 ppb TMO chromatographic peak (top) and mass spectrum of the TMO peak (bottom)

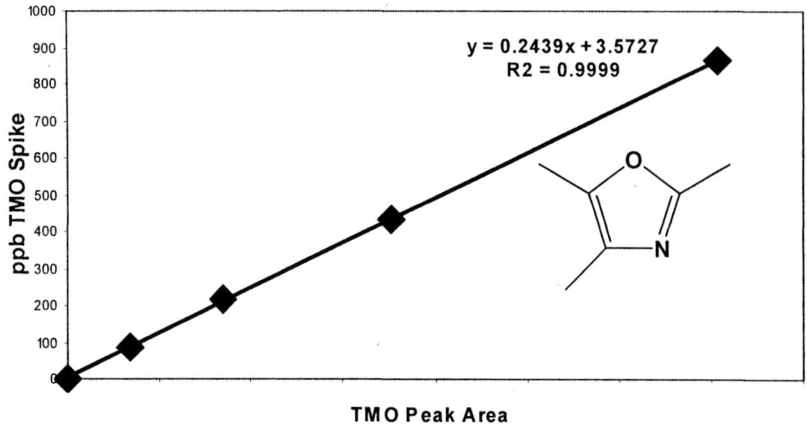

Figure 2. Calibration curve for TMO-spiked sour cream powder.

variability was unknown; however, dimethyl sulfide (DMS) was considered the possible flavor impact chemical because of reports in the literature describing DMS as the cause of a creamed-corn flavor in beer (*6,7*). While DMS and its creamed-corn flavor are undesirable in the context of beer, in cheese products the flavor is regarded as favorable, increasing the savory character of cheese.

Because of the high volatility of dimethyl sulfide, the SPME method was modified to include the use of a 75 μm carboxen/poly(dimethylsiloxane) fiber instead of the poly(dimethylsiloxane)/divinyl benzene fiber. This method (Method 1) provided inadequate precision and a poor linear correlation coefficient for standard curves. The method was modified by lowering preincubation and extraction temperatures (Method 2).

- Method 1
  - 4 min preincubation @50°C; 20 min extraction @50°C.
  - 75 μm carboxen/poly(dimethylsiloxane) fiber.
  - EMS internal standard (200 ppb spike in ethanol).
  - Peak areas by extracted ion mode (*m/z* 60).
- Method 2
  - Same as Method 1 but preincubation and extraction temperatures are 25°C and extraction time is only 15 min.

As shown in Table I, milder extraction conditions resulted in higher linear correlation coefficients for standard curves, lower relative standard deviations of replicate analyses and improved sensitivity for dimethyl sulfide in cheese powder when analyzed by Method 2 compared to Method 1.

Table I. Comparison of Experimental Parameters for Two SPME Methods for DMS in Cheese Powder

Experimental Parameter	*Method 1*	*Method 2*
Correlation coefficient[a]	0.670	0.992
Internal standard	Ethyl methyl sulfide	Ethyl methyl sulfide
Average peak area of replicates[b]	25340±26%	67628±4.5%

[a]Linear least squares correlation coefficient for standard curve by method of additions of spiked samples.

[b]Average peak area of five replicate analyses (±standard deviation).

Figure 3 illustrates the correlation between creamed-corn flavor and concentration of dimethyl sulfide. An optimum creamed-corn flavor was observed in cheese powders with a dimethyl sulfide concentration of approximately 400 ppb. Levels lower than 400 ppb were associated with weaker creamed-corn flavor; higher levels could result in off-flavors associated with rotten vegetables. Interestingly, the addition of dimethyl sulfide to product formulation prior to spray drying was found to be an effective way to enhance the creamed-corn flavor attribute in spray-dried cheese powders.

Off-flavors in a Club Cheese Product

Unacceptable off-flavor problems were occurring with a club cheese product used in cheese powder formulations. The customer of the cheese powder, however, insisted the cheese powder manufacturer use this particular club cheese because of the unique earthy/mushroom flavor notes it contributed to the powder.

Based on descriptive sensory analysis, eight club cheese samples were found to have acceptable flavor, three samples had borderline flavor acceptability and three samples had unacceptable flavor. Samples with unacceptable flavor profiles were criticized as having a dirty, musty and/or sour taste. SPME GC-MS analysis using the general methodology was used to profile acceptable, borderline and unacceptable samples. On average, approximately 45 different organic volatiles were determined in each sample.

Multivariate analysis (Pirouette®, Ver. 3.11, Infometrix, Inc., Woodinville, WA) was applied to the data set to determine if the SPME method was capable of extracting/detecting the chemicals responsible for causing the dirty, musty, sour flavor defect. Two-dimensional PCA results, illustrated in Figure 4, show that clusters for acceptable, borderline and unacceptable products are generated from plots of data, indicating that the chemicals responsible for the off-flavor

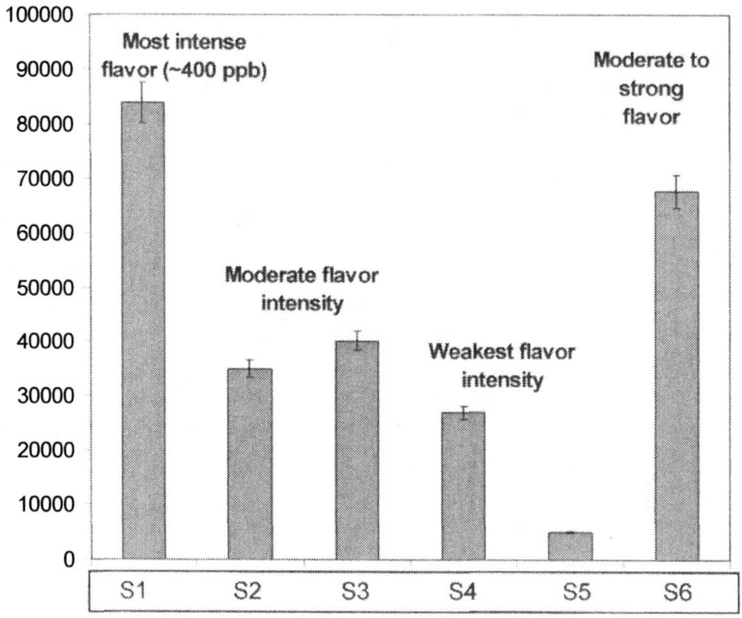

Figure 3. DMS peak areas for six samples of cheese powder with varying intensities of creamed-corn flavor. Error bars represent standard deviation of four replicate analyses by SPME GC-MS.

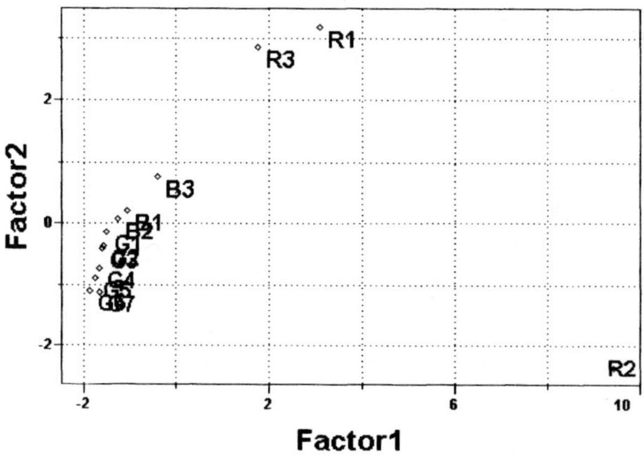

Figure 4. Two-dimensional PCA plot of club cheese samples with acceptable flavor (G), borderline flavor (B) and unacceptable flavor (R).

defect are being monitored by the SPME method. Table II lists the chemicals with the highest modeling power in constructing the PCA plot. Several of these compounds, including methyl ketones, 2-heptanol and 8-nonen-2-one, were recognized as possible mold metabolites.

Samples of this particular type of club cheese were observed to develop extensive surface mold growth after one week storage at 4.4°C. Samples of surface mold were scraped from these samples and analyzed by SPME GC-MS. The major volatiles in surface molded cheese scrapings are shown in Table III. Several of these volatiles were the same ones that were shown to have high modeling power in the PCA plot. Based on these results, it was determined that the desirable earthy flavor notes that the cheese powder customer wanted were attributed to mold growth. When mold growth was too extensive, however, the dirty, musty, sour off-flavors became severe, leading to product rejection.

Significant reduction in off-flavor complaints was achieved when the cheese powder manufacturer demanded that the club cheese supplier maintain tighter Q.C. controls for mold growth and limit rework.

Table II. Ten Volatiles With Highest Modeling Power for PCA Clustering Shown for Club Cheese Samples in Figure 4

Rank	*Chemical*	*Modeling Power*
1	Isopropyl octanoate	0.968
2	2-Undecanone	0.958
3	8-Nonen-2-one	0.957
4	2-Nonanone	0.956
5	2-Propanol	0.943
6	Nonanal	0.943
7	Amyl acetate	0.940
8	2-Heptanone	0.923
9	2-Heptanol	0.921
10	2-Methyl-1-butanol	0.915

Off-Flavor in Cheese Powders

Over a period of several years, samples of one type of cheese powder occasionally failed sensory flavor tests. The cheese powder manufacturer wanted to learn what chemicals were responsible for sample rejections.

SPME GC-MS analysis using the typical method was conducted on a set of 13 samples that passed sensory testing, nine samples that failed and a gold standard (i.e., a sample that the customer selected as having optimum flavor

Table III. Volatiles Present as Major Peaks in Chromatogram of Mold Scrapings from Club Cheese

Chemical	*Flavor Descriptor*	*Chemical*	*Flavor Descriptor*
p-Methyl anisole	Dirty/musty	Isoamyl alcohol	Vinous
1-Oceten-3-ol	Mushroom	1-Pentanol	Vinous
8-Nonen-2-one[a,b]	Fatty/earth	Isopropyl octanoate[a]	Fruity
2-Heptanol[a]	Mushroom	Amyl acetate[a]	Sweetish
Isobutyl alcohol	Vinous	2- and 3-Octanone	Soapy/resinous

[a]High modeling power in PCA plots.
[b]Significant flavor component of blue and mold-ripened cheese.

characteristics). Approximately 60 different volatiles were identified, and 34 of these volatiles were considered to be possible contributors to flavor. Multivariate analysis (Pirouette) was again used to identify compounds most likely causing rejection. Figure 5 shows PCA clustering of pass and fail samples based on peak areas of the 34 chemicals identified by SPME GC-MS as possible contributors to flavor. Inspection of PCA loading plots of the 34 chemicals and results of modeling power statistics showed that organic acid peak areas were the most significant drivers for clustering of pass and fail samples.

To test the rationality of organic acids being significant contributors to cheese powder flavor, odor units and taste units were calculated for a typical acceptable cheese powder sample. Organic acids were quantitated by SPME GC-MS using the extracted ion at m/z 60. Pentanoic acid was used as an internal standard at a spike level of 350 ppm. Odor and taste threshold concentrations for organic acids were obtained from the literature.

As shown in Table IV, levels of organic acids in the cheese powders are 10 to 1000 times higher than their odor and taste thresholds, making these compounds important contributors to the flavor of cheese powder.

To further test the importance of organic acid levels to cheese powder flavor, 25 additional cheese powders were analyzed by SPME GC-MS. This group of samples was previously subjected to sensory analysis. Using Pirouette's K-nearest neighbor (KNN) algorithm, a pass/fail prediction model was created from the peak area results of the original data set. The prediction model was then used to predict the pass/fail status of the 25 additional cheese powders. Based only on organic acid peak areas, the model was able to predict seven of nine fail samples correctly and 15 of 16 pass samples correctly, for an overall prediction rate accuracy of 88%.

Maintaining consistent ratios of organic acids in cheese powder ingredients was shown to be an important way to consistently optimize the flavor of the finished product.

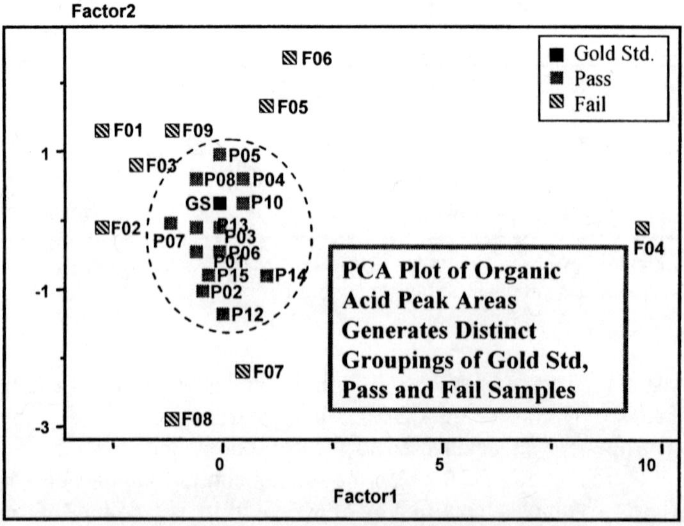

Figure 5. PCA plot of pass, fail and gold standard cheese samples based on organic acid peak area data.

Table IV. Concentrations, Odor Unit Values and Taste Unit Values of Organic Acids in a Typical Cheese Powder

Organic Acid	*Concentration in Cheese Powder (ppm)[a]*	*Threshold in Water (ppm)*		*Log U_o[b]*	*Log U_t[c]*
		Odor	*Taste*		
Butyric	700	0.24	6.6	3.5	2.0
Isovaleric	50	0.12	----	2.6	----
Hexanoic	500	3.0	5.4	2.2	2.0
Heptanoic	30	3.0	----	1.0	----
Octanoic	225	3.0	5.3	1.7	1.6
Decanoic	300	10.0	3.5	1.5	1.9

[a]Concentration determined by SPME GC-MS; pentanoic as an internal standard.

[b]U_o = Volatile concentration/odor threshold in water.

[c]U_t = Volatile concentration/taste threshold in water.

SPME GC-TOFMS Advantages

To adequately define GC peaks, either qualitatively or quantitatively, it is preferable to have 20-30 data points (or spectra) across the peak. Scanning mass spectrometers like the Agilent MSD typically acquire data at a maximum rate of one to ten spectra per second, which is insufficient for some applications. In contrast, the time-of-flight mass spectrometer (TOFMS) like the Pegasus III (Leco Corp., St. Joseph, MI) is an array detector and does not scan; rather it measures all of the ions across the *m/z* range simultaneously and has data acquisitions rates ranging from a few spectra per second to hundreds of spectra per second (8).

Complementary to the speed advantage is the spectral reproducibility of TOFMS. Unlike slow, scanning mass spectrometers, where the sample concentration changes that occur in the source during elution of a chromatographic peak cause distortion of the mass spectrum, TOFMS is a "snapshot" technique, where ion packets are extracted and mass analyzed almost simultaneously. This results in unskewed mass spectra. The combination of unskewed mass spectra and peak definition offered by TOFMS rapid acquisition rates allow for the application of powerful peak-find and deconvolution algorithms.

When SPME GC-TOFMS is used for flavor profiling of cheese-based products, our lab typically detects approximately 25% more extracted compounds in samples because of the application of deconvolution algorithms. The additional peak information gained can be quite useful in elucidating flavor impact chemicals in cheese products. Figure 6 shows how a dimethyl sulfone peak "buried" in a large butyric acid peak in a chromatogram of a cheese powder sample was detected and accurately quantified by TOFMS.

Conclusion

SPME GC-MS for the flavor analysis of cheese-based products offers numerous benefits. The technique is:

- Rapid.
- Relatively inexpensive to perform.
- Extracts a wide range of volatiles/semivolatiles.
- Sensitive (ppb).
- Accurate & reproducible.
- Versatile.
- Automatable.
- Works well with ancillary techniques such as multivariate analysis, TOFMS and GC-olfactometry.

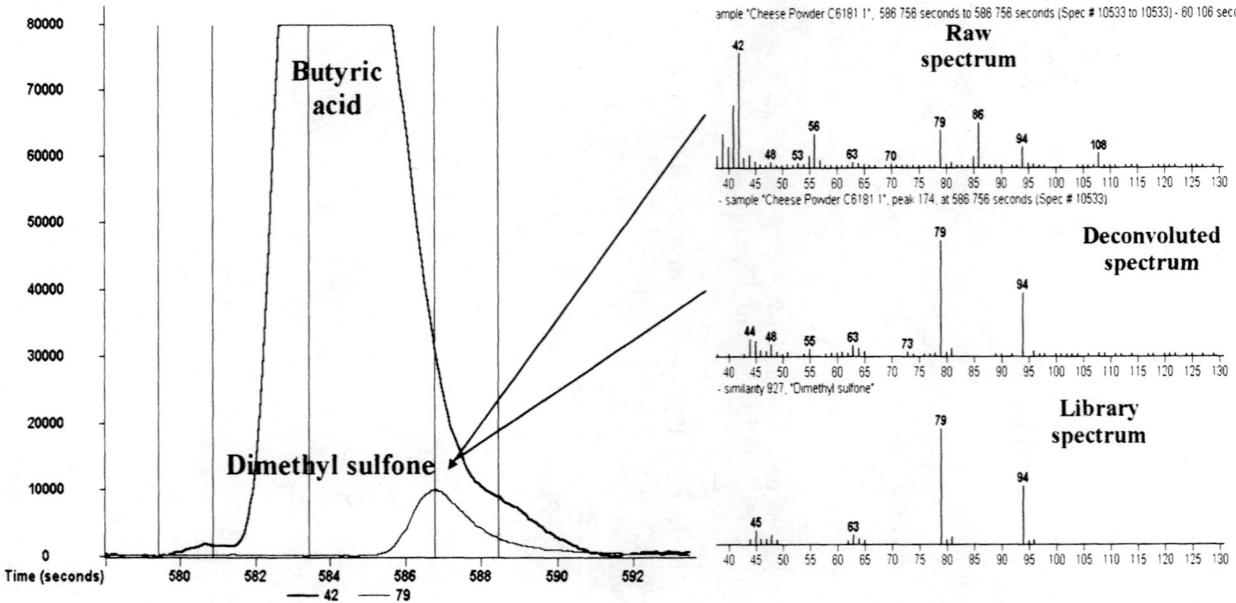

Figure 6. SPME GC-TOFM chromatogram of cheese powder from 580 to 592 seconds showing coelution of butyric acid (m/z *42) and dimethyl sulfone* (m/z *79). Also shown are the raw, deconvoluted and library spectra of the dimethyl sulfone peak. (See page 1 of color inserts.)*

References

1. Werkhoff, P.; Brennecke, S.; Bretschneider, W.; Bertram, H.-J. Modern methods for isolating and quantifying volatile flavor and fragrance compounds, in *Flavor, Fragrance and Odor Analysis*; Marsili, R.T., Ed.; Marcel Dekker, Inc., New York, 2002, p 139.
2. Lee, J.-H.; Diono, R.; Kim, G.-Y.; Min, D.B. Optimization of solid-phase microextraction for the headspace volatile compounds of Parmesan cheese. *J. Agric. Food Chem.* **2003**, *51*, 1136-1140.
3. Anocibar Beloqui, A. Contribution to the study of sulfur compounds of the red wines. Ph.D. Thesis Number 6111, University Victor Segalen Bordeaux 2, Bordeqeaux, France, 1998.
4. Griffith, R.; Hamond, G. Generation of Swiss cheese flavor components by the reaction of amino acids with carbonyl compounds. *J. Dairy Sci.* **1988**, *72*, 604-613.
5. Pripis-Nicolau, L.; de Revel, G.; Bertrand, A.; Maujean, A. Formation of flavor components by the reaction of amino acid and carbonyl compounds in mild conditions. *J. Agric. Food Chem.* **2000**, *48*, 3761-3766.
6. Anderson, R.J.; Clapperton, J.F.; Crabb, D.; Hudson, J.R. Dimethyl sulfide as a feature of lager flavor. *J. Inst. Brew.* **1975**, *81*, 208-213.
7. Aness, B.J.; Bamforth, C.W. Dimethyl sulfide—A Review. *J. Inst. Brew.* **1982**, *88*, 244-252.
8. Holland, J.F.; Bardner, B.G. The advantages of GC-TOFMS for flavor and fragrance analysis, in *Flavor, Fragrance and Odor Analysis*; Marsili, R.T., Ed.; Marcel Dekker, Inc., New York, 2002, p 107.

Chapter 5

The Influence of Texture on Aroma Release and Perception Related to Dairy Products

Jennifer Mei and Gary Reineccius*

Department of Food Science and Nutrition, University of Minnesota, 1334 Eckles Avenue, Saint Paul, MN 55108

Aroma release studies from foods during eating have considered both the direct effects of food aroma component interaction and the indirect effects of food texture. Direct interactions typically are the result of specific chemical binding/interaction between some food constituent and specific aroma components. Textural effects consider how much of a barrier food structure presents to aroma release during eating and is potentially much broader in influence. In this chapter, we present an overview of the literature on how food texture influences aroma release from foods/model systems in general and from yogurt and cheese as examples of dairy products.

There has been substantial interest in understanding how texture may influence the liking of a food. One can readily acknowledge that texture itself is a key attribute influencing liking. However there is also a body of knowledge that suggests that texture influences other sensory inputs that may also drive liking, e.g. taste or aroma perception. Understanding texture-flavor interactions is difficult because both flavor and texture are problematic to measure. Whether measured sensorially or instrumentally, there has been a great deal of disagreement on the subject and a slow evolution in knowledge over time. This review will discuss some of the research related to texture-aroma interactions in model and food systems. Unfortunately, there is little information on the role of texture in influencing flavor perception in dairy products, the topic of the symposium this paper is based upon, and thus we will deal more generally with the subject and provide some additional discussion of the limited studies on dairy products.

Definition of Food Texture

It is beneficial to initially define texture. Lawless and Heymann (*1*) have defined texture as the "characteristics of a product perceived by the visual or tactile senses including geometric qualities, surface attributes, perceived changes under deforming forces or when forced to flow if liquid, phase change behavior such as melting and residual tactile sensations following chewing, swallowing and expectoration". This definition alludes to the complexity of defining and measuring texture. Texture involves many senses including sight, sound and touch (*1*). In essence, texture is an attribute of a food product as a result of its structure (in the broadest sense) and behavior during handling and eating (*2*).

Texture perception is complex and varies with individuals. Part of the reason for this is that manipulation or deformation of the product is necessary for perception to occur. The forces, rates, and other parameters of this manipulation will change the textural sensations from a physiological and physical perspective. Additionally, texture perception results from combining tactile and kinesthetic receptor inputs that are genetically determined (*3*).

Perception of texture begins before the food enters the mouth and continues past swallowing. Anticipation based on previous experience starts the process. During this stage, visual cues of size, shape, and viscosity are considered (*2*). This leads to manipulation of the food either through handling or using utensils. The way a piece of meat cuts or the sound a potato chip makes when broken are examples of the textural information we perceive before the food enters the mouth (*2*). Touch is part of the texture perceived when the food first enters the mouth. Movements of the tongue increase the shear rates and provide information regarding elasticity, stickiness, and viscosity (*2*). Crushing and grinding in the beginning of mastication provide fracturing information (*2, 3*). During the first few chews the consumer experiences, crispness, brittleness, and plasticity (*4*). Subsequent chewing provides information on particle size, consistence and adhesion (*2*). After swallowing, mouthfeel and residual food are evaluated (*2*).

Aroma:texture Interactions - Sensory Analysis

Over the years, sensory scientists have approached aroma:texture interactions using methods such as descriptive analysis, time-intensity ratings. One of the earlier studies (*5*) reported on the influence of five different hydrocolloids on the perception of four aroma compounds. While the hydrocolloids generally decreased aroma intensity, the effect varied with the aroma compound and hydrocolloid rather than absolute viscosity (*5*). This observation suggests that viscosity is not the controlling factor, i.e. limits surface area exposed on eating or volatile diffusion from the bulk of the food to the air:food interface, but a specific interaction between a given hydrocolloid and an aroma compound. In a later study Pangborn *et al.* (*6*) applied these findings to commercial products (orange drink, tomato juice, and coffee). In these products, aroma intensity was suppressed when hydrocolloids were added. Of the five hydrocolloids studied (xanthan gum, hydroxypropylcellulose, sodium alginate, and medium and low viscosity carboxymethylcellulose), xanthan gum reduced aroma intensity to the greatest extent across the three beverage types (*6*).

Studies on gels yielded similar results: hydrocolloids reduced the perceived intensity of aroma compounds. For example, the maximum intensities of benzaldehyde, d-limonene, and ethyl butyrate were lower in firm gelatin and carrageenan gels than in soft or medium gels (*7*). In agreement, Kalviainen *et al.* (*8*) studied high viscosity gels which mimic the texture of candies and found that the aroma of gels with weak and fragile textures were rated more intense than those of firmer gels. Gel texture clearly influences the perceived intensity of food aroma.

Consistent with the observations that hydrocolloid type influences the intensities of perceived aroma compounds in viscous solutions, aroma intensity in gels is similarly affected by the gelling agent used. Aroma intensity for given aroma compounds and concentrations is greater from starch and gelatin gels than carrageenan gels (*7*). Pectin gels have faster release and higher aroma intensities than gelatin, starch or gelatin/starch gels (*8*).

As the food system becomes more complex, the role of aroma-texture interactions is less clear. Yogurt is a food product where texture contributes greatly to consumer acceptability. The addition of thickener (pectin/starch mixture) to low fat stirred yogurt was found to suppress green apple flavor but did not influence other flavor notes studied (*9*). It was proposed that an interaction between hexenal (green apple note) and the dairy proteins contributed to this observation. Other proposed hypotheses included: 1) stimulation by sweetener rather than volatiles caused the decreased green apple intensity; or 2) a perceptual interaction between the volatile and the yogurt viscosity occurs (*9*, *10*). Conversely, thickness perception decreased as flavor concentrations increased (*10*). Additional work ruled out that this effect is due to dilution by the flavor, but did not exclude trigeminal effects (*10*). In a study of 120 young people (11-14 years old), thickness of yogurt did not affect the flavor intensity (*11*).

Saint-Eve *et al.* (*12*) provided evidence that the aroma characteristics and number of aromas present will affect texture perception. They found that coconut and butter aromas led to a thicker perception of low fat yogurt than green apple and almond which were considered smoother (*12*). Mouth coating and smoothness went up and thickness and stickiness went down in yogurts flavored with an aroma mixture as opposed to a single aroma compound. Since the yogurts flavored with multiple aromas had a higher total aroma intensity, it is possible that the decrease in perceived thickness was due to the higher aroma concentration (*10*, *12*). An alternative explanation was that the aroma complexity affected texture perception on a cognitive level (*12*).

Measuring aroma-texture interactions sensorially is a substantial challenge because aroma and texture perceptions are individual experiences: different people experience different sensations. However, there is a trend that aroma-texture interactions change with age. Age specific texture by flavor interactions were documented using white cream soups (*13*). Kremer *et al.* (*13*) found that thickening soup with potato starch caused an increase in perceived thickness of both mushroom and unflavored soups in young panelists (18-29 years old) whereas perceived thickness only increased in unflavored soups in elderly panelists (60-84 years old). More relevant to this paper is that increasing the viscosity of mushroom flavored soups did not change the perceived flavor intensity for the elder panelists but *increased* it for the younger panelists (*13*). The observation that flavor intensity increased with thickener addition is inconsistent with nearly all of the previous literature and is rather impossible to reconcile with any theory other than cognitive effects.

Age is not the only human parameter contributing to differences in perceived arome-texture interactions: other factors include mastication and chewing efficiency. Chewing efficiency affects the way people experience aroma and texture during eating (*14*). Time-intensity scaling found that temporal flavor perception patterns are linked to mastication patterns (*15*).

Overall, sensory studies show that thickening and gelling agents have the ability to suppress flavor intensity and that gelling agent/aroma combinations are important in determining the intensity of aroma perception. Additionally, we see that individual variation can affect the way people perceive flavor and texture and their interactions.

Aroma-texture Interactions using Instrumental Analysis

Since sensory studies have documented reduced intensities of aroma perception when the solution is either increased in viscosity or forms a gel, instrumental analysis has been employed to see if these perceived differences are due to the amount or timing of aroma release during eating. There has been a generally accepted notion that viscous solutions (or gels) reduce aroma release through less exposed surface area of the food in the mouth and/or reduced diffusion through the viscous solution (or gel) to the food/air interface. The advent of analytical tools to study the *in-vivo* release of aroma compounds from foods during eating has provided the opportunity to investigate these hypotheses.

Investigating several model systems, Bylaite *et al.* (*16*) and Hansson *et al.* (*17*) both concluded that viscosity of the model system did not affect aroma release as measured by equilibrium or dynamic headspace analysis methods. In more detail, Hansson *et al.* (*17*) found that pectin concentration (0.5 – 2.5%) had no effect on the equilibrium headspace concentrations of isopenthyl acetate, cis-3-hexenyl acetate, l-menthone, ethyl hexanoate, or linalool. Limonene in the headspace decreased only at the 2.5% pectin level. It should be noted for later reference that when similar viscosities were obtained using pectin and simple sugars, the effects were very different: the pectin either had no effect (five of the six model aroma compounds) or decreased headspace concentrations (limonene) while sugars (>20% concentration) generally increased headspace concentrations of volatiles (except limonene). Hansson *et al.* (*17*) hypothesized that the water activity of the solution was a controlling factor of volatile release (Figure 1). This is an interesting hypothesis in that other authors have often cited "salting out" as being responsible for increases in volatile release (when observed). Very often these authors were working with relatively dilute sugar or salt solutions, or hydrocolloid levels that could not be responsible for binding sufficient water to affect "salting out". Hansson *et al.* (*17*) presented data that showed no "salting out" effects until glucose concentrations exceeded 20%, and sucrose or glucose syrup levels exceeded 40%. Few foods other than confectionery products contain such high sugar levels. Hansson *et al.* (*17*) make a very concise concluding statement that "...viscosity has no effect on flavour release ...". It must be noted that Hansson *et al.* (*17*) used an equilibrium method to measure aroma release and thus viscosity influences related to mass transfer would not be observed.

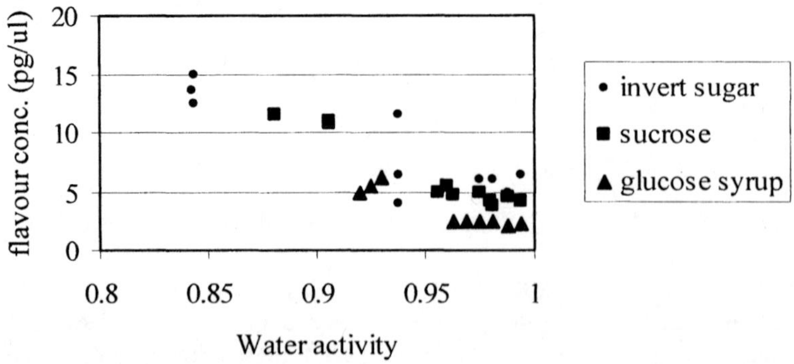

Figure 1. Effect of water activity on the release of cis-3-hexenol from model systems (Reproduced with permission from reference 17. Copyright 2001 Elsevier.)

Bylaite *et al.* (*16*) used a dynamic headspace method (as opposed to an equilibrium method) to conduct a similar study as Hansson *et al.* (*17*). While equilibrium headspace methods have been used for many years to determine aroma interactions with food constituents, it has been argued that equilibrium headspace methods do not consider mass transfer related effects on aroma release. For example, a food ingredient that slows diffusion of a volatile in a food will not influence the equilibrium headspace of that volatile but would slow its release under dynamic conditions. Dynamic headspace methods are considered to be more representative of aroma release in the mouth.

Bylaite *et al.* (*17*) found that pectin addition and resultant changes in solution viscosity did not affect the release of model aldehydes even under dynamic conditions. In a later study, Bylaite *et al.* (*18*) found that aroma suppression in λ-carrageenan solutions did not correspond to changes in viscosity. Once again, their study concluded that viscosity changes in the matrix were not important to aroma release: matrix composition was considered the dominant factor (*17*).

Using pectin, locust bean gum, starch, or guar gum as fat replacers in fat-free stirred yogurt, Decourcelle *et al.* (*19*) illustrated that different thickeners have different effects on aroma release (dynamic method). They hypothesized that some polysaccharides reduced aroma release from yogurt because of polysaccharide interactions with the aroma compounds. Locust bean gum increased aroma release and guar gum showed no effect. This increase was attributed to alterations in water activity and a salting-out effect. The suggestion that salting out is responsible for this observation is highly improbable. A concentration of 0.1g locust bean gum/100g of fruit preparation would not bind sufficient water to affect any salting out of volatiles. Previous authors have repeatedly shown that in excess of 20% monosaccharides are required to bind sufficient water to affect a salting out of volatiles. The concept that any food ingredient at low concentrations (where it does not significantly reduce the free water) can increase the volatility of aroma compounds is untenable unless it results in a co-distillation of the volatile. One can conceive that an escaping gas (e.g. CO_2, N_2 or perhaps ethanol) might carry a volatile into the product headspace, but it is inconceivable that a non-volatile, e.g. artificial sweetener or hydrocolloid, in the ppm or low % can have such an effect. One has to question the reliability of the data when this effect is observed or offer a new theory

Decourcelle *et al.* (*19*) also examined rheological parameters and shear conditions for their samples. Their conclusion was that thickeners have differing effects on aroma release due to their chemical makeup as opposed to their rheological parameters. As will be discussed, Seuvre *et al.* (*20*) agreed with this conclusion.

Seuvre *et al.* (*20*) studied aroma release from model systems with different structures (single phase liquid, emulsion and gel) and compositions (water, milk, β-lactoglobulin and β-lactoglobulin plus medium chain triglycerides (Miglyol)). They found a minor effect of physical form (gel versus non-gelled system) and a major effect of matrix composition and nature of the volatile compounds. It is relevant that they produced gelled systems through heating. In most other studies gelling has been induced by changes in system composition which confounds data interpretation, i.e. were effects due to changes in composition or physical state.

Preparing equal viscosity solutions, Roberts *et al.* (*21*) also showed that different thickeners (carboxymethylcellulose (CMC), guar gum and sucrose) influenced release differently (dynamic release). Guar and CMC had very similar effects on the release of all model aroma compounds (variable magnitude of decrease) while sucrose reduced the release of certain model volatiles (α-pinene, ethyl-2-methylbutyrate and 1,8-cineole) to a much greater extent (Figure 2). In this case it is very interesting that sucrose resulted in *decreased* release since the addition of sugars has been associated with salting out effects: that is, increases in volatile release. Furthermore, the volatiles that were influenced the greatest were the apolar compounds. Salting out effects have generally been reported for the more water soluble volatiles observing no effect on apolar compounds. This would be rational since the binding of water would have the greatest effect on those compounds that are water soluble. However, occasionally there have been reports that volatile release from sucrose solutions was reduced for apolar volatiles. It was hypothesized that sucrose makes an aqueous solution less polar thereby making it a better solvent for apolar volatiles (reducing their release). Thus, there appears to be an opportunity for further studies to better define the effect of sucrose on aroma release of volatiles.

There is no question that the physical/chemical properties of aroma compounds have a strong influence on their release properties from a food. As

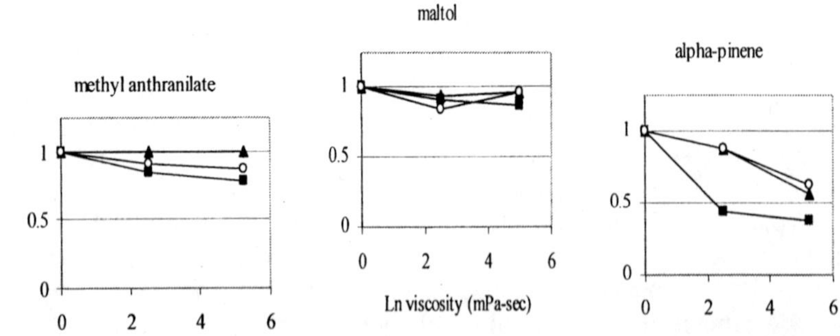

Figure 2. Effects of sucrose (■), guar gum (▲) and carboxymethyl cellulose (O) on the release of volatile flavor compounds under dynamic conditions. (Y axis – Imax normalized to water) (Reproduced from reference 21. Copyright 1996 American Chemical Society.)

an example, Taylor *et al.* (*22*) used an API-MS to study *in-vivo* release of a range of volatiles from mixed phase gel systems. They found that both matrix and volatile compound have an influence on release and that the aroma compound is a primary factor influencing release. Additionally, persistence of aroma in the breath after drinking an aqueous solution is most affected by the volatile compound. Further examination using a quantitative structure:property relationship approach found that major terms in the model express the hydrophobicity and vapor pressure of the aroma molecule (*23*).

Correlating Sensory and Instrumental Analysis

The disparity between sensory and instrumental studies on aroma-texture interactions is apparent. Sensory studies repeatedly show that viscosity affects aroma intensity while instrumental results indicate that viscosity is not an important parameter in aroma release: the aroma compound properties and texture agent affect aroma release in instrumental studies. A direct comparison of the literature on the sensory and instrumental studies discussed previously is not possible because of the different methods and systems used in the studies. A better understanding of this paradox can be obtained by reviewing studies which considered release from both a sensory and instrumental perspective at the same time.

Simultaneous sensory time-intensity analysis and nosespace aroma release by API-MS show that aroma *perception* is affected by texture but *in-vivo* aroma release is not (at least as measured by Imax). Looking at model dairy desserts, aroma intensity measured by sensory panel varied with texture (Figure 3 top). When the nosespace results from these samples were examined,

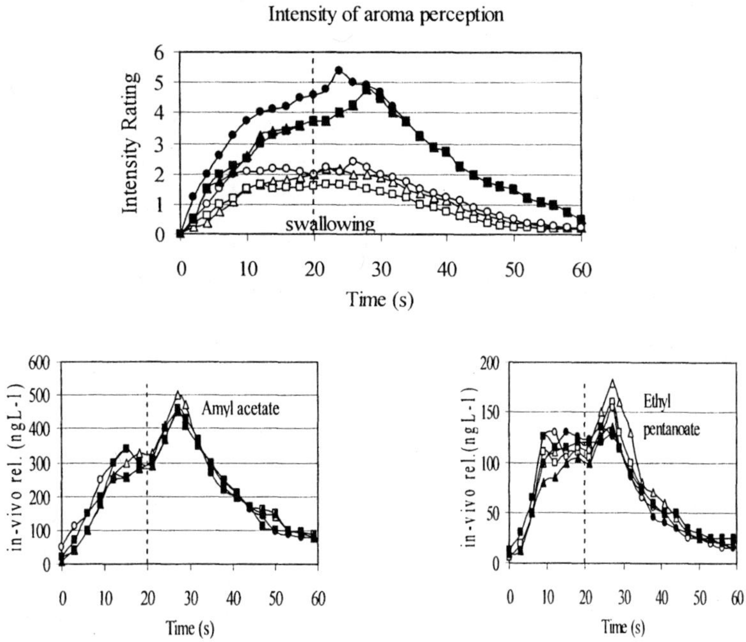

Figure 3. The effect of viscosity (obtained by varying sucrose concentration and carageenan type) on flavor intensity (top), and on the in-vivo release of two aroma compounds (I max, bottom right and left; □,Δ,○: 25gkg^{-1}; ■,▲,● 100gkg^{-1}) (Adapated from reference 24. Copyright 2004 American Chemical Society.)

aroma release concentrations showed insignificant effects of texture (*24*). The model desserts in this study were formulated with sugar and as a result aroma-texture interactions may have been confounded by aroma-sweetness interactions.

To address this concern, Weel *et al.* (*25*) studied unsweetened whey protein gels and found nosespace concentrations (Imax) are independent of gel hardness and water binding capacity (Figure 4). Both of these studies show that while texture affects perceived aroma intensity, it does not significantly influence nosespace aroma concentrations (*24*, *25*). Based on this observation, it was assumed that texture influences on flavor perception are the result of factor(s) other than volatile or taste release. It was proposed that this may be a learned response and be due solely to a cognitive function.

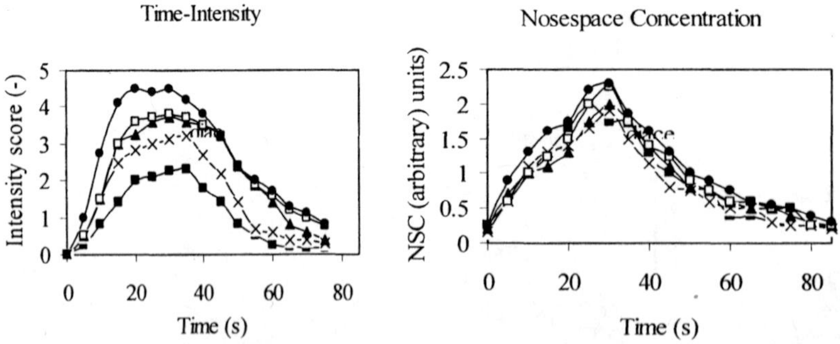

Figure 4. Averaged time-intensity sensory recordings (left) and diacetyl release profiles (right; NSC = nose space concentration) from gels of varying strengths (increasing gel strength → ●, □, ▲, X, ■,) (Reproduced from reference 25. Copyright 2002 American Chemical Society.)

If nosespace aroma concentration (Imax) does not vary with texture and sensory perception does, it goes to reason that some parameter(s) other than nosespace aroma concentration is (are) responsible for the sensory aroma-texture interactions observed. One alternative is that much of the work discussed thus far has considered only the effect of system viscosity on the Imax (maximum intensity) of a volatile in the nosespace. Imax may not be correlated with sensory perception of aroma intensity but other nosespace parameters of aroma release may. In a study of sucrose-gelatin gels, it was shown that maximum nosespace concentration did not change with gelatin concentration while sensory intensity did (as expected). However, there was a correlation between the *rates* of volatile release and sensory response (*26*). Mestres *et al.* (*27*, *28*) have followed up on this idea in recent publications. Like other authors they found that Imax, area under the curve, and Tmax were not significantly different between soft and hard whey protein gels in *in-vivo* testing. However, they found that preswallowing data differed for these gels. While the olfactory system does not receive any odorants when eating viscous foods or very soft gels until swallowed (the air passage from the mouth to the olfactory region is blocked by the velum), at some gel strength the velum is opened by chewing and the individual receives olfactory inputs before swallowing: in essence, the olfactory receptors get bursts of aroma for sensing during chewing. In Figure 5, one can see that during the chewing process, model aroma compound is released and that the model aroma compound is released more quickly (reaches Imax more rapidly) from the soft gel than the hard gel. Mestres *et al.* (*27*, *28*) hypothesized that since the aroma bursts provided by chewing a soft gel increase in volatile concentration more rapidly than chewing a hard gel, the impact on perception is more intense. This is based on the concept that if a sensory input increases slowly (e.g. chewing a hard gel) it will be perceived as being less intense than if the input increases rapidly (soft gel) even if both stimuli reach the same maximum intensity. They use the term "First Impression" to describe this hypothsis.

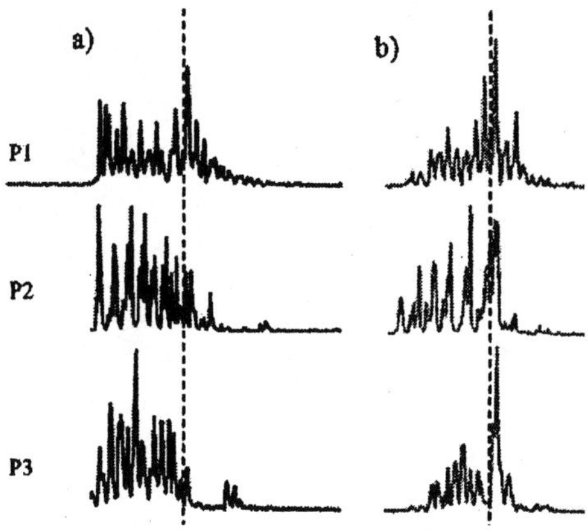

Figure 5. Selected ion trace for m/z 117 from PTR nose space of the consumption of (a) soft gel (4% protein) and (b) hard gel (10% protein) for three different panelists. The dashed line indicates the swallow breath (Adapted from reference 27. Copyright 2005 American Chemical Society.)

Work reported by Cook *et al.* (*30*) is supportive of the "First Impression" hypothesis but does not rule out cognitive effects. In this work, oral shear stress was found to correlate well with both aroma and sweetness perception in viscous solutions (*30*). Since oral shear stress would indicate how the food was masticated (i.e. at low shear stresses all foods would be eaten with a closed velum to prevent choking but are eaten with an open velum at higher shear stresses), one would expect to see a relationship as shown in Figure 6. One would need to relate oral shear stress to eating pattern (open Vs. closed velum) to better interpret these data.

Eljade and Kokini (*31*) originally showed that oral shear stress, which is predicted by physical properties of viscous foods, correlates to perceived viscosity in the mouth.

Cook *et al.* (*29*, *30*) explored the possibility that texture may influence the release of tastants which would influence aroma perception. They found that perceived saltiness influences savory aroma perception in viscous solutions while nosespace aroma concentration (Imax) is unchanged. Similar results were observed with sweet flavors (*9*). There is little question that any parameter that influences the delivery of tastants normally associated with a given flavor is going to influence overall flavor perception. Unfortunately, there does not

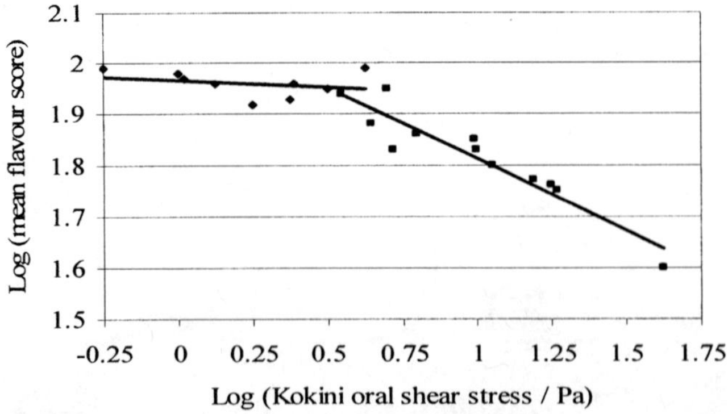

Figure 6. Relationship between perceived intensity of isoamyl acetate in a model system and the oral shear stress of the model system. (Reproduced with permission from reference 30. Copyright 2003 Oxford University Press.)

appear to be any definitive data determining if system viscosity actually reduces the delivery of tastants to *taste receptors*. (Granted there are data monitoring tastant release from model systems into the saliva during eating) There is ample sensory data in the literature to shown taste perception decreases as system viscosity increases but we do not absolutely know if this is due to reduced delivery to taste receptors or reduced perception (cognitive effect).

Understanding Aroma:texture Interactions

Sensory studies generally agree that aroma-texture interactions occur: in particular, increasing texture (viscosity or gel strength) decreases aroma intensity. There is also no question that the food matrix composition (i.e. fat level, thickener, protein level, or sweetner types) and aroma compounds (i.e. volatility, chemical structure) are primarily responsible for variations in aroma release (which may influence flavor intensity and/or character).

There is some uncertainity in understanding how viscosity (or gel strength) affects flavor intensity. Substantial work has suggested that texture does not influence aroma release and thus, differences in flavor intensity across systems of differening viscosity are due to either taste-flavor or texture-cognitive effects. However, recent work of Mestres *et al.* (*27*, *28*) demonstrated that texture influences how a food is masticated (position in the mouth and open Vs. closed velum) and the aroma release rate, the combined effects of which are correlated to perceived intensity. This work suggests that aroma release is a determent of flavor intensity. This does not mean that taste interactions or cognitive effects are irrelevant but perhaps less important than originally thought. It is clear that a multidisciplinary approach is necessary and fortunately growing more common in order to better understand aroma release and flavor perception (*30*).

Texture and Flavor Perception in Dairy Products – Yogurt

Texture-flavor perception is relevant to food products for at least two reasons. First, if there is a flavor issue in these products (e.g. why isn't the flavor of a low fat cheese perceived to be the same as a full fat cheese?), one must know the source of the problem. It would make little sense to try to change the aroma profile of a low fat cheese if, in fact, the issue is one of texture changing the perceived flavor. Certainly low fat cheeses differ in texture from their full fat counterparts, and aged cheeses differ from young cheeses, texture may be a factor in this flavor difference. As another example, in yogurts, numerous hydrocolloids are used to provide viscosity and physical stability to the product. Each hydrocolloid will change the texture of the product, aroma release and likely, flavor perception. It is relevant to understand the effect of these different hydrocolloids on the flavor of the product. While there are a very limited number of papers focused on dairy products involving texture:flavor interactions, we will provide a discussion of the literature available starting with yogurt.

Of the dairy products, there is more research on yogurt flavor-texture relationships than any other dairy product. As mentioned above, yogurts generally contain added hydrocolloids to provide water binding and textural properties. Some of the work already presented in this paper has dealt with yogurt model hydrocolloid systems and provided relevant data. We will now look specifically at work on yogurt as a food.

Brennan *et al.* (*11*) reported on the influence of flavor on texture perception, texture on flavor perception, and color on both texture and flavor perception as well as liking in yogurt systems. They conducted only sensory testing and worked with 11-14 year olds. They varied texture by using two different wheat starches. Contrary to expectations based on the basic literature discussed above, they did not find a significant influence of flavor on texture, or texture on flavor (measured intensity). This may be due to the panel composition (120 untrained, young consumers) or that the difference in viscosity between the two yogurts was not adequate to elicit a significant response (viscosity range or magnitude of change). As discussed above, Cook *et al.* (*30*) found that viscosity effects were not detectable at low viscosities but became more evident at higher viscosities (observation supported by Mestres (*27*, *28*)).

Paci Kora *et al.* (*10*) also worked with complete yogurt formulations noting that little work studying texture:flavor interactions had been done on complex systems (as opposed to model systems). They varied the viscosities of their yogurts by using two levels of hydrocolloid (starch and pectin blend) and through mechanical shear after fermentation. They evaluated textural effects on perception (attribute profiling) using a trained sensory panel (10 members). The primary findings were that flavor concentration and type influenced texture perception (consistent with other researchers, e.g. Saint-Eve *et al.* (*12*)) and texture decreased the intensity of some flavor attributes notably green apple aroma and sweet taste (had no significant effect on the other 11 sensory attributes). They noted that the decrease in green apple aroma was not explained by a measurable chemical interaction between the hydrocolloids and the primary volatile responsible for this sensory note (hexanal). Thus, they postulated that this effect may be due to binding by the whey protein or be a cognitive effect based on viscosity.

Decourcelle *et al.* (*33*) have reported on the effect of numerous thickeners and sweeteners on the release of model aroma compounds from stirred, fat-free yogurts (discussed earlier in this paper). This study did not include sensory evaluation but instrumental analysis of aroma release under dynamic conditions. As previously discussed in detail, they found pectin and starch to reduce aroma release, guar and sweeteners to have no effect, and locust bean gum to increase release.

Mei *et al.* (*34*) studied *in-vivo* aroma release from complete yogurt formulations using Proton Transfer Reaction Mass Spectrometry. They found the release of the ethyl butanoate, (Z)-hex-3-enol, and ethyl 3-methylbutanoate to be suppressed slightly by sweeteners, with 55 DE high-fructose corn syrup having the greatest effect. Addition of thickening agents had no significant effect on the aroma release profiles of the compounds under study. No sensory data were gathered in this study.

From the very limited amount of research done on yogurts, a primary conclusion is that aroma release from yogurt is not significantly influenced by textural changes. This may be the result of not montoring the relevant release parameters, e.g. preswallow data, or the changes in viscosity in yogurt are too small to influence aroma release. Thus, while numerous studies have shown that the use of different hydrocolloids, or varied hydrocolloid levels, affect sensory perception of flavor, the relevance of this observation to yogurt is not clear. In particular instances when the use of a thickener (hydrocolloid) has an effect on flavor release (and perhaps perception), it is likely due to unique binding of some component(s) of the flavoring to a given thickener. In reviewing the sensory data, it is of particular interest that textural effects on yogurt flavor were not readily observed in a large scale consumer test (11). One then can ask how important texture-flavor interactions are within the viscosity ranges occurring in commercial products.

Texture and Flavor Perception in Dairy Products – Cheese

There has been nothing published determining specifically the influence of cheese texture on flavor perception using combined sensory and instrumental analyses. This is a particularly problematic task in cheese since this type of study requires the ability to independently vary texture and flavor of the samples for study. However, there has been instrumental work on aroma release with some correlations to time-intensity data. A brief overview of some of this work will be presented.

Delahunty *et al.* have published numerous papers in this general area many of which have been on methods of measuring aroma release. One of the earlier papers they have published on flavor perception in cheese evaluated the utility of two instrumental methods (buccal headspace and temporal buccal headspace) to predict sensory attributes of cheese (*35*, *36*). They had some success in relating the instrumental data to the sensory data. Unfortunately, the subject of texture-flavor interactions (topic of this manuscript) has not been explored by this group.

Pionnier *et al.* (*37*, *38*) have considered flavor release from cheeses in various ways. The earlier paper reported on the release of tastants during eating processed cheese. They used Atmospheric Pressure Ionization Mass Spectrometry (API-MS) to monitor the release of model tastants (amino acids, acids and minerals) into saliva during eating. Their primary findings were that the release parameters were more dependent upon human physiology than the physiochemical properties of the tastants. This is not necessarily true of aroma release during eating where volatility plays a major role in determining release. The authors did not consider texture in their study.

Pionnier *et al.* (*38*) went on to better characterize the oral parameters that influenced the release of volatiles from a model cheese during eating. They used API-MS and off-line Solid Phase Micro Extraction-Gas Chromatography-Mass Spectrometry (SPME-GC-MS) as instrumental methods to monitor volatile release. No sensory analysis was used in this study. The oral parameters monitored reflected breathing, chewing, swallowing, mastication and salivation. Their summary statements indicated that mastication and breathing parameters were best related to aroma release. Again, no textural considerations were included in this study.

Salles *et al.* (*39*, *40*) chose to select three soft cheeses (camembert (pasteurized and unpasteurized milk) and Brie) and study the release of intrinsic volatiles and non-volatiles (tastants) from these cheeses by sensory (temporal profiling) and instrumental methods (*in-vivo*). The release of volatiles, tastants and perception were influenced by cheese type but there was no textural work to relate release or perception to textural properties. The only significant correlation between *in-vivo* aroma release and perception was for a sulfury note.

Differences in release between cheeses was attributed to cheese compositional factors.

There is little currently in the literature that has considered texture-flavor interactions in cheese. Thus the data on yogurt are the most complete and useful to us in determining the role of texture in influencing flavor perception in dairy products.

Literature Cited

1. Lawless, H.T.; Heymann, H.H. *Sensory Evaluation of Food: Principles and Practices;* Chapman & Hall: New York, 1998.
2. *Food Texture: Measurement and Perception;* Rosenthal, A.J., Ed.; Aspen Publishers: Gaithersburg, MD, 1999; pp. 1-17
3. Christensen, C.M. *Advan. Food Research* **1984**, *29*, 159-199.
4. Sherman, P. J. Food Sci. **1969**, *34*, 458-462.
5. Pangborn, R.M.; Szczesniak, A. *J. Texture Studies* **1974**, *4*, 467-482.
6. Pangborn, R.M.; Gibbs, Z.M.; Tassan, C. *J. Texture Studies* **1978**, *9*, 415-436.
7. Guinard, J.X.; Marty, C. *J. Food Sci.* **1995**, *60*, 727-730.
8. Kalviainen, N.; Roininen, K.; Tuorila, H. J. *Texture Studies* **2000**, *31*, 407-420.
9. Hollowood, T.A.; Linforth, R.S.T.; Taylor, A.J. *Chem. Senses* **2002**, *27*, 583-591.
10. Paci Kora, E., Brossard, C. ; Rousseau, F. ; Llamas, G.; Dumont, J.-P. *J. Sensory Studies* **2003**, *18*, 367-390.
11. Brennan, M., Setser, C.; Schmidt, K.A. *J. Food Sci.* **2002**, *67*, 2785-2789.
12. Saint-Eve, A., Paci Kora, E.; Martin, N. *Food Qual. Pref.* **2004**, *15*, 655-668.
13. Kremer, S., Mojet, J.; Kroeze, J.H.A. *J. Texture Studies* **2005**, *36*, 255-272.
14. Brown, W.E., Dauchel, C.; Wakeling, I. *J. Texture Studies* **1996**, *27*, 433-450.
15. Wilson, C.E.; Brown, W.E. *J. Sensory Studies* **1997**, *21*, 69-86.
16. Bylaite, E.M.; Meyer, A.E.; Adler-Nissen, A.S.J. *J. Agric. Food Chem.* **2003**, *51*, 8020-8026.
17. Hansson, A.; Andersson, J.; Leufven, A. *Food Chem.* **2001**, *72*, 363-368.
18. Bylaite, E. ; Ilgunaite, Z.; Meyer, A.S. ; Adler-Nissen, J. *J. Agric. Food Chem.* **2004**, *52*, 3542-3549.
19. Decourcelle, N.; Lubbers, S.; Vallet, N.; Rondeau, P.; Guichard, E. *Internat. Dairy J.* **2004**, *14*, 783-789.
20. Seuvre, A-M; Espinosa Diaz, M.; Cayot, P.; Voilley, A. *Lait* **2004**, *84*, 305-316.

21. Roberts, D.D.; Elmore, J.S.; Langley, K.R.; Bakker, J. *J. Agric. Food Chem.* **1996**, *44*, 1321-1326.
22. Taylor, A.J.; Besnard, S.; Puaud, M.; Linforth, R.S.T. *Biomol. Engin.* **2001**, *17*, 143-150.
23. Linforth, R.S.T.; Taylor, A.J. *J. Agric. Food Chem.* **2000**, *48*, 5419-5423.
24. Lethauaut, L.; Weel, K.G.C.; Boelrijk,; Alexandra, E.M.; Brossard, C.D. *J. Agric. Food Chem.* **2004**, *52*, 3478-3485.
25. Weel, K.G.C.; Boelrijk, A.E.M.; Alting, A.C.; van Mil, P.J.J.M.; Burger, J.J.; Gruppen, H.; Voragen, A.G.J.; Smit, G. *J. Agric. Food Chem.* **2002**, *50*, 5149-5155.
26. Baek, I.; Linforth, R.S.T.; Blake, A; Taylor, A. *J. Chem. Senses* **1999**, *24*, 155-160.
27. Mestres, M.; Moran, N.; Jordan, A.; Buettner, A. *J. Agric. Food Chem.* **2005**, *53(2)*, 403-409.
28. Mestres, M.; Kieffer, R.; Buettner, A. *J. Agric. Food Chem.* **2006**, *54*, 1814-1821.
29. Cook, D.J.; Linforth, R.S.T.; Taylor, A.J. *J. Agric. Food Chem.* **2003**, *51*, 3067-3072.
30. Cook, D.J. ; Hollowood, T.A.; Linforth, R.S.T.; Taylor, A.J. *Chem. Senses* **2003**, *28*, 11-23.
31. Elejalde, C.C.; Kokini, J.L. *J. Texture Studies* **1992**, *23*, 315-336.
32. *Flavor Perception;* Taylor, A.J.; Roberts, D.D. Eds.; Blackwell Publishing Ltd: Oxford, UK, 2004; pp. 283.
33. Paci Kora, E.; Latrille, E.; Souchon, I.; Martin, N.J. *Sensory Studies* **2003**, *18*, 367-390.
34. Decourcelle, N.L.;Vallet, N.; Rondeau, P.; Guichard, E. *Internat. Dairy J.* **2004**, *14*, 783-789.
35. Mei, J.B.; Reineccius, G.A.; Knighton, W.B.; Grimsrud, E.P. *J. Agric. Food Chem.* **2004**, *52*, 6267-6270.
36. Delahunty, C.M.; Piggott, J.R.; Conner, J.M.; Paterson, A. *Italian J. Food Sci.* **1996**, *8(2)*, 89-98.
37. Delahunty, C.M.; Piggott, J.R.; Conner, J.M.; Paterson, A. *J. Sci. Food Agric.* **1996**, *71*, 273-281.
38. Pionnier, E.S.; Salles, C.; Le Quere, J-L; Taylor, A.J. In *Flavour Research at the Dawn of the Twenty-First Century*, Le Quere, J.L.; Etievant, P.X., Eds., Lavoisier and Intercept: Londres-Paris-New York, 2003; pp. 248-250.
39. Pionnier, E.C.; Chabanet, C.; Mioche, L.; LeQuere, J.L. ; Salles, C. *J. Agric. Food Chem.* **2004**, *52*, 557-564.
40. Salles, C.H.; Holowood, T.A.; Linforth, R.S.T.; Taylor, A.J. In *Flavour Research at the Dawn of the Twenty-First Century*, Le Quere, J.L.; Etievant,

P.X., Eds., Lavoisier and Intercept: Londres-Paris-New York, 2003; pp. 170-175.

41. Salles, C.E.; Engel, E.; Nichlaus, S.; Taylor, A.J.; Le Quere, J.L. In *Process and Reaction Flavors*, Weerasinghe, D.K.; Sucan, M.K., Eds., ACS Symposium Series 905. American Chemical Society: Washington, DC, 2005; pp. 192-207.

Chapter 6

Streamlined Analysis of Short-, Medium-, and Long-Chain Free Fatty Acids in Dairy Products

Keith R. Cadwallader[1], Tanoj. K. Singh[2], and John P. Jerrell[1]

[1]Department of Food Science and Human Nutrition, University of Illinois, 1302 West Pennsylvania Avenue, Urbana, IL 61801
[2]Food Science Australia, 671 Sneydes Road, Private Bag 16, Werribee, Victoria 3030, Australia

In dairy products free fatty acids (FFAs) with fewer than ten carbon atoms can impart both desirable and undesirable (rancid) flavor notes, while long-chain FFAs are responsible for a soapy taste. The determination of FFAs in milk, cheese and other dairy products is particularly complicated owing to two main factors, i.e. FFAs (2 to 20 carbon atoms) represent <0.5% of total fat and many of the short-chain FFAs are extremely volatile. The best approach is to analyze short-chain FFAs (<C10) separately from the long-chain FFAs. Short-chain FFA can be extracted in soap form with water and then upon acidification of the extract short chain FFAs can be extracted into ether for GC analysis. Long chain FFA can be extracted with a mixture of ether:heptane (1:1, v/v), resulting in an extract that contains the whole range of lipids including FFAs. The present paper will describe a streamlined approach (utilizing an aminopropyl bonded phase column) for the analysis of total FFAs in dairy products.

Free fatty acids (FFAs) are present in very small amounts in most biological systems and are often not quantified as a separate lipid class. However, careful quantification of FFAs is important in certain cases, such as in milk and dairy products where they can impact flavor. Depending on their relative abundance, short-chain FFAs with fewer than eight or ten carbon atoms can impart both desirable and undesirable rancid flavor notes to food products, while long-chain fatty acids are primarily responsible for a soapy taste (Table I). The pH also has an influence on the flavor of FFAs. For example, at the pH of Cheddar cheese (pH ~5.2) a considerable portion of the FFAs are present as salts, which are nonvolatile and of less flavor significance (*1*).

Table I. Predominant Fatty Acids in Dairy Products

Fatty acid		*Aroma/Taste Property*
Butanoic acid (butryric)	C4:0	Fecal, cheesy
Hexanoic acid (caproic)	C6:0	Sweaty, goat-like
Octanoic acid (capryllic)	C8:0	Body odor, sweaty
Decanoic acid (capric)	C10:0	Waxy, rancid, fatty
Dodecanoic acid (lauric)	C12:0	Waxy, soapy
Tetradecanoic acid (myristic)	C14:0	Soapy
Hexadecanoic acid (palmitic)	C16:0	Soapy
Octadecanoic acid (stearic)	C18:0	Soapy
9-Octadecenoic acid (oleic)	C18:1	Soapy
9, 12-Octadecadienoic (linoleic) acid	C18:2	Soapy
9, 12, 15-Octadecatrienoic (linolenic) acid	C18:3	Soapy

Data from Jensen et al. (*2*)

Various methods have been developed for the analysis of FFAs. It is possible to estimate total fat acidity by use of colorimetric or titration methods, but no information about individual FFAs is provided by these methods. Gas chromatography (GC) is the most common method used for determination of volatile short-chain FFAs (i.e. C_2 to C_{10} FFAs) (*3*). These FFAs can be measured as free acids since they are volatile and stable enough for GC. Static headspace and dynamic headspace GC techniques have been applied for the analysis of butanoic acid and other short-chain FFAs (*3, 4, 5*). It is also possible to perform GC analysis after an aqueous extraction of the relatively polar (soap form) short-chain FFAs (*6*). Short-chain FFAs may also be measured by high performance liquid chromatography (HPLC)(7). However, with HPLC the FFAs must still be extracted and precolumn derivitization or fluorescence detection is generally required to achieve high enough sensitivity (*7*). Capillary zone electrophoresis has also been applied for the analysis of short-chain FFAs (*8*).

In contrast to the volatile short-chain FFAs, the medium- to long-chain FFAs are not volatile enough for direct analysis by GC. Therefore, they are most often determined after derivatization into more volatile and stable methyl esters. The isolation of minor amounts of these FFAs from a complex sample containing mainly triacylglycerides (TAGs) has been the most challenging aspect of this analysis. Kaluzny et al. (*9*) demonstrated that aminopropyl bonded phase columns could be used to separate FFAs from the bulk lipid material with high yield and purity. Other researchers have attempted to avoid additional sample preparation by performing derivatization of the crude extract within the injection port of the GC (i.e. use of tetramethylammonium hydroxide for methylation of FFAs), but this approach was found to be unsuitable for samples with high TAG to FFA ratios (*10*). Therefore, it is necessary to isolate the medium- to long-chain FFAs from the abundant TAGs and other lipids prior to derivatization and GC analysis.

The GC determination of FFAs in milk, cheese and other dairy products is particulary complicated owing to three main factors: (i) FFAs represent only <0.5% of the total fat, which is primarily composed of triacylglycerols (TAGs), (ii) FFAs with chain lengths from 2 to 20 carbon atoms are often present and (iii) short-chain FFAs with fewer than eight caron atoms are volatile and subject to loss during sample preparation; whereas FFAs with more than ten carbon atoms require separation and a derivatization step prior to GC analysis. The present paper describes a stream-lined sample preparation method for the detailed GC analysis of short-, medium- and long-chain FFAs in dairy products.

Experimental Procedures

Materials

Samples (5 in total) used in this study were commercial enzyme modified cheeses. Samples were stored away from light at -20° C in amber glass jars equipped with Teflon® closures.

Analytical grade diethyl ether, heptane, chloroform, methanol, borontrifloride (BF_3, 14% w/v in methanol), 2-ethylbutanoic acid, sodium chloride, and ahydrous sodium sulfate were obtained from Sigma-Aldrich, Inc. (St. Louis, MO). Standard FFAs listed in Tables II and III and heptadecanoic acid were purchased from Nu Chek Prep (Elysian, MN). Bond Elute aminopropyl disposable columns (500 mg) were obtained from Varian, Inc. (Palo Alto, CA). Odor-free water was prepared by boiling deionized-distilled water in an open flask until its volume was decreased by one-third.

Sample Preparation

Short-Chain FFAs

The method used here was modified from Innocente et al. (*6*).

- Homogenize 5 g of cheese, plus 25 mL of deodorized water and 10 μL of internal standard (0.976 μg/μL of 2-ethylbutanoic acid in methanol).
- Centrifuge mixture at 3000 x g for 10 min.
- Remove 5 mL aliquot of supernatant (pH 5 – 6) and acidify to pH ~ 2 using aqueous 10% w/v HCl).
- Add 2 g of NaCl and extract with 2 mL of diethyl ether.
- Dry ether layer over 2 g sodium sulfate.
- Extract ready for GC analysis.

Medium- and Long-Chain FFAs

Sample preparation for analysis of medium- and long-chain FFAs involved three distinct steps: (i) solvent extraction of fat from cheese was adapted from Chavarri et al. (*10*), (ii) the method of Kaluzny et al. (*9*) was adapted for the separation of FFAs from bulk lipid classes and (iii) fatty acid methyl esters (FAMES) were prepared with BF_3-methanol as described in AOCS official method Ce 2-66 (*11*).

<u>Step 1: Fat extraction</u>

- Grind 5 g of cheese with 5 g of anhydrous sodium sulfate.
- Mix with 3 mL of aqueous 2.5M H_2SO_4 plus 10 μL of internal standards (0.89 μg /μL of heptadecanoic acid).
- Extract mixture with 5 mL of diethyl ether-heptane (1:1 v/v).
- Centrifuge (3000xg, 10 min) and repeat extraction (2X).
- Combined organic fractions containing crude fat + FFAs.

<u>Step 2: Separation of free fatty acids from bulk lipid classes</u>

- Apply the combined organic phase (from above) to an aminopropyl-bonded phase column, previously equilibrated with 10 mL of heptane.
- The **triacylglyerides** (TAGs) are eluted with 10 mL of chloroform/2-propanol (2:1 v/v).
- **Free fatty acids** (FFAs) are eluted with 5 mL of 2% (v/v) formic acid in diethyl ether.
- Eluate contains FFAs

<u>Step 3: Preparation of methyl esters of separated free fatty acids</u>

- Evaporate eluate (from above) to dryness under nitrogen stream at ambient temperature.
- Add 2 mL of BF_3-methanol to residue.

- Mix and boil mixture for 2 min.
- Add 2 mL of hexane.
- Mix and boil mixture in sealed tube for 1 min.
- Collect hexane layer containing **free fatty acid methyl esters** (FAMES).
- Dry over 1 g anhydrous sodium sulfate
- Extract ready for GC analysis.

Analysis of Extracts

Gas Chromatography

The GC system consisted of an HP5890 series II (Agilent Technologies, Inc., Palo Alto, CA) equipped with a split/splitless injector and flame ionization detector (250°C). For analysis of short-chain FFAs, extracts were injected (2 μL) in the splitless mode (250°C, 30 s valve delay time) into a DB-FFAP (15 m x 0.53 mm i.d. x 1.0 μm film; J&W Scientific, Folsom, CA) column. Helium was the carrier gas at a constant flow of 5 mL/min. GC oven temperature was programmed from 35 to 225°C at a rate of 8°C with intial and final hold times of 5 and 30 min, respectively. For analysis of FAMES, extracts were injected in the split mode (250°, 1:50 split ratio) into a DB-WAX (30 m x 0.25 mm i.d. x 0.25 μm film; J&W Scientific) column. Helium flow was the carrier gas at a constant flow of 1 mL/min. GC oven temperature was programmed from 35 to 170°C at a rate of 10°C/min, then to 230° at a rate of 6°C/min, with intial and final hold times of 5 and 30 min, respectively.

Standard curves for the short chain acids for quantification were created by using pure individual fatty acids at three levels in 5 mL of water, in place of sample, and then the above procedures were followed. Standard curve for the medium to long chain acids was created using a commercial FAME mixture in hexane.

Recovery Studies

Recovery test was performed in duplicate, by adding known quantities of short chain (C2-C10) and medium to long chain (C10-C18) fatty acids to an enzyme modified cheese (sample 1).

Results and Discussion

Extraction and Analysis of Short-chain FFAs

The gas chromatographic analysis of total free fatty acids (FFAs) in dairy products such as cheese requires a two step approach to minimize the interference from the bulk lipid, which is primarily composed of triaceylglyerides. Efficient and reproducible isolation of short-chain FFAs can be achieved by taking advantage of their moderate water solubilities in the pH range of 5 to 6. Accurate quantitation can be achieved by use of an internal standard which is added to the sample prior to extraction. In the present study, 2-ethylbutanoic acid was chosen as an appropriate internal standard since it's polarity and volatility fall within the range of short-chain FFAs to be measured. Furthermore, it is well resolved from the target FFAs by GC (Figure 1) and has not been previously reported as a volatile FFA component of dairy products.

The approach taken in the present study is capable of determining FFAs from C_2 to C_8 with good recovery and excellent precision of less than 5% for nearly all FFAs determined (Table II). An exception was decanoic acid which was recovered at only about 80% due to its poor solubility in water. This FFA was more effectively determined using the method developmed for medium and long-chain FFAs.

Normal chain FFAs were found in all samples. Some branched acids, e.g. methylpropanoic acid and 3-methylbutanoic acid, were present at low levels. FFAs such as butanoic, pentanoic, hexanoic, octanoic and decanoic acids originate from milk fat, but acetic/propanoic, methylpropanoic and 3-methylbutanoic acid originate from metabolism, e.g. during ripening of cheese, from lactose/citrate, valine and leucine, respectively (*1*).

Extraction, Separation and Analysis of Medium- and Long-chain FFAs

In contrast to the volatile short-chain FFAs, the medium- and long-chain FFAs are nonvolatile and relatively nonpolar in nature, making it difficult to separate them from the highly abundant TAGs. The use of a disposable aminopropyl bonded phase column allowed for the efficient isolation of the FFAs from the other nonpolar lipids (e.g. TAGs) (Figure 2, Table III). The medium- and long-chain FFAs were nearly completely recovered and coefficient of variation was less than 3% for all FFAs. The high degree of accuracy and excellent precision of this method can be largely attributed to the use of an internal standard (heptadecanoic acid) since the various sample extraction, workup and derivatization steps could potentially lead to considerable error in the analysis.

Need for Comprehensive FFA Analysis

Streamlined and comprehensive FFA analysis of enzyme modified cheeses, as well as other dairy products, has potential for the evaluation of both aroma- and taste-active FFAs. FFAs in Table II would primarily impact the overall

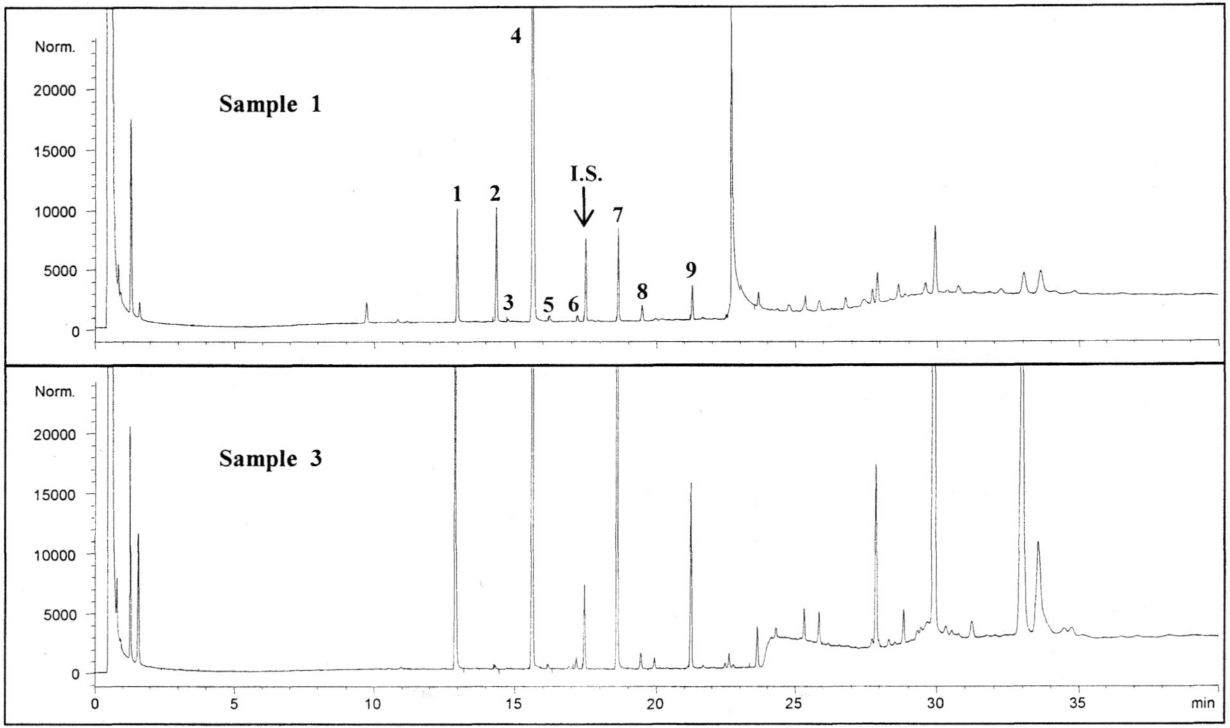

Figure 1. Gas chromatogram of short-chain fatty acids in samples 1 and 3.
[1 = acetic acid, 2 = propanoic acid,3 = methylpropanoic acid, 4 = butanoic acid, 5 = 3-methylbutanoic acid, 6 = pentanoic acid, 7 = hexanoic acid, 8 = octanoic acid, 9 = decanoic acid, I.S. = internal standard]

Table II. Composition of Short-chain Free Fatty Acids (FFAs) in Enzyme Modified Cheeses

			Sample 1		*Sample 2*		*Sample 3*		*Sample 4*		*Sample 5*	
FFA	*%Rec*[a]	*%CV*[b]	*mg/Kg*[c]	*%CV*	*mg/Kg*	*%CV*	*mg/Kg*	*%CV*	*mg/Kg*	*%CV*	*mg/Kg*	*%CV*
Acetic	99	4.8	318	3.0	441	2.7	1870	3.3	4510	3.5	201	2.5
Propionic	119	2.7	61.4	1.7	9.16	0.8	7.95	1.3	7.74	0.9	7.44	1.5
Methylpropionic	112	1.3	0.57	1.3	0.35	2.3	0.42	1.9	0.32	2.5	ND[d]	ND
Butanoic	109	1.5	182	0.8	24.7	1.2	294	1.1	423	2.1	392	1.7
3-Methylbutanoic	103	2.5	1.31	0.2	1.18	0.2	1.08	0.9	2.28	1.5	ND	ND
Pentanoic	101	0.9	2.28	1.8	1.61	1.3	3.91	1.8	5.09	2.2	4.77	2.1
Hexanoic	99	1.1	18.0	1.5	11.5	0.6	189	1.1	268	1.3	216	0.8
Octanoic	96	2.3	5.04	2.1	3.52	2.1	46.8	2.1	195	2.1	87.9	2.1
Decanoic (C10:0)	*81*	*5.6*	*189*	*3.8*	*145*	*3.8*	*523*	*3.8*	*638*	*3.8*	*51.2*	*3.8*

[a]Percent recovery. [b]Percent coefficient of variation (n = 2), [c]Average concentration (mg/kg) wet sample basis (n = 2), [d]ND = not detected.

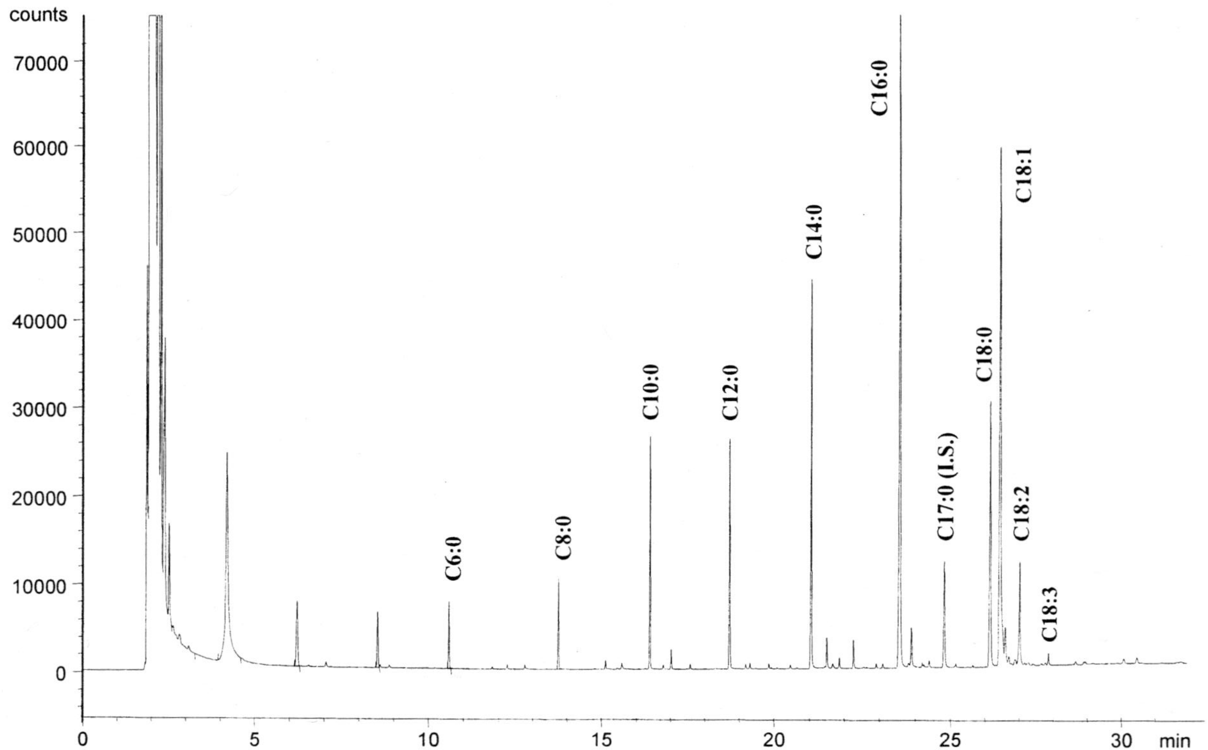

Figure 2. Gas chromatogram of medium- and long-chain fatty acids methyl esters (sample 1).

Table III. Composition of Medium- and Long-chain Free Fatty Acids (FFAs) in Enzyme Modified Cheeses

			Sample 1		*Sample 2*		*Sample 3*		*Sample 4*		*Sample 5*	
FFA	*%Rec*[a]	*%CV*[b]	*mg/Kg*[c]	*%CV*	*mg/Kg*	*%CV*	*mg/Kg*	*%CV*	*mg/Kg*	*%CV*	*mg/Kg*	*%CV*
C10:0	101	2.1	175	2.3	169	2.6	1650	1.7	1490	2.8	355	1.5
C12:0	111	1.5	174	1.7	165	2.1	1200	1.6	724	2.4	236	1.9
C14:0	107	1.2	425	1.4	378	1.6	2800	1.6	1310	1.6	692	1.9
C16:0	105	1.7	1470	1.2	1460	1.1	8300	1.1	3960	1.7	3780	2.5
C18:0	103	1.4	899	1.2	1060	1.3	4220	1.4	3080	1.9	3699	1.4
C18:1	106	1.1	2070	1.4	2570	1.4	2550	1.3	2800	1.3	3940	1.5
C18:2	99	2.2	383	1.9	485	1.9	558	1.8	590	1.6	610	2.1
C18:3	103	2.5	40.5	2.1	56.5	1.9	81.3	2.2	93.4	1.9	94.4	2.4

[a]Percent recovery. [b]Percent coefficient of variation (n = 2), [c]Average concentration (mg/kg) wet sample basis (n = 2).

aroma (or odor) of a product; whereas, those FFA in Table III would impart an undesirable soapy taste. Some FFAs, such as octanoic and decanoic acid, could potentially impart both aroma and taste to the product. The ability to assess the impact of both potentially positive aroma-impact FFAs and potentially negative taste-active FFAs should be of particular importance during the assessment of enzyme modified cheeses.

FFAs are involved in several types of reactions which vary in importance with the type of dairy product involved (see reference *1* for details). Methyl ketones are produced from fatty acids by oxidative degradation. Methyl ketones are responsible for the characteristic aroma of blue-veined cheeses. Another reaction in which polyunsaturated and, perhaps, monounsaturated, fatty acids can be involved, is oxidation. The extent of oxidation in cheese is, however, rather limited, possibly due to a low redox potential. Aliphatic and aromatic esters play an important part in the flavor and, sometimes, the off-flavor of cheese. This synthesis mainly concerns the above mentioned short- or medium-chain fatty acids and the alcohols involved may be aliphatic (ethanol), aromatic (phenylethanol) or thiols (methanethiol). Esters can be produced enzymatically, by lactic acid bacteria but can also easily result from a purely chemical reaction. Esters generally contribute a fruity flavor to dairy products which is desirable and characteristic in many cheeses (Parmesan, Parrano) but undesirable in others (Cheddar). γ- and δ-Lactones have been identified in cheeses, particularly in Cheddar, where they have been considered as important for flavor. Lactones are cyclic esters resulting from the intramolecular esterification of hydroxy acids through the loss of water to form a ring structure. Lactones possess a strong aroma, which although not specifically cheese-like, may be important in the overall cheese flavor impact.

Conclusions

The two methods decribed here for the streamlined analysis of short-, medium- and long-chain FFAs are straight forward to use and result in high yields and excellent precision for all predominant dairy FFAs. It is necessary to separately analyze the more volatile short-chain FFAs away from the nonvolatile medium- and long-chain FFAs, due to potential loss of the former FFAs during sample workup and derivatization. A crucial step in the analysis of medium- and long-chain FFAs is the separation of FFAs from the bulk of the nonpolar lipid components. Use of internal standards in both methods is necessary to assure accuracy and repeatability due the number of steps involved in the assays.

Literature Cited

1. Singh, T.K.; Drake, M.A.; Cadwallader, K.R. *Comp. Rev. Food Sci. Food Safety* **2003**, *2*, 139-162.
2. Jensen, R.G.; Ferris, A.M.; Lammi-Keffe, C.J. *J. Dairy Sci.* **1991**, *74*, 3228-3243.
3. Ulberth, F. *Z. Lebensm. Unters. Forsch A* **1998**, *206*, 305-307.
4. Woo, A.H.; Kollodge, S.; Lindsay, R.C. *J. Dairy Sci.* **1984**, *67*, 874-878.
5. Chen, L.; Boyle-Roden, E.; Rankin, S.A. 1999. *J. Food Sci.* **1999**, *64*, 390-392.
6. Innocente, N.; Moret, S.; Conte, L.S. *J. Agric. Food Chem.* **2000**, *48*, 3321-3323.
7. Totsuka, S.; Matsuda, H.; Sato, T.; Minagawa, N. *Bunseki Kagaku* **1996**, *45(10)*, 927-932.
8. Buchberger, W.; Winna, K. *Mikrochemica Acta* **1996**, *122(1-2)*, 45-52.
9. Kaluzny, M.A.; Duncan, L.A.; Merritt, M.V.; Epps, D.E. *J. Lipid Res.* **1985**, *26*, 135-140.
10. Chavarri, F.; Virto, M.; Martin, C.; Najera, A.I.; Santisteban, A.; Barron, L.J.R.; De Renobales, M. *J. Dairy Res.* **1997**, *64* 445-452.
11. AOCS. Official methods and recommended practices of the American Oil Chemist Society, 4th Edition (eds. Mehlenbacher V.C.; Hopper, T. H.; Sallee, E. M.; Link, W. E.; Walker, R. O.; Walker, R. C.; Firestone, D.), **1989**, Method Ce 2-66.

Chapter 7

From Pasture to Cheese: Changes in Terpene Composition

Hedwig Schlichtherle-Cerny[1,*], Miroslava I. Imhof[1], Estrella Fernández García[2], and Jacques O. Bosset[1]

[1]Agroscope Liebefeld-Posieux (ALP), Swiss Federal Research Station for Animal Production and Dairy Products, Berne, Switzerland
[2]Instituto Nacional de Investigación y Tecnología Agraria y Alimentaria (INIA), Madrid, Spain

The monoterpene composition of fresh grass originating from altitudes between 1400 and 1920 m in Switzerland, of milk stemming from the same sites, and of a Raclette-type cheese which had been produced from the same milk were analyzed by dynamic headspace-GC-MS. The milk and cheese samples revealed additional monoterpenoids which had not been detected in the grass, such as 3,7-dimethyl-1,6-octadiene, 3,7-dimethyl-2-octene, 2,6-dimethyl-2,6-octadiene and menth-1-ene. To study the impact of rumen fermentation on the monoterpenes, a model plant, cow parsnip (*Heracleum sphondylium* L.), was incubated with rumen fluid for 24 h. All terpenes were partially degraded, β-myrcene, (*E*)- and (*Z*)-β-ocimene almost completely. The degradation of the terpenes was correlated with the formation of the additional new compounds detected in milk and cheese. These compounds may be formed by hydrogenation during rumen fermentation.

Introduction

Terpenes are secondary natural plant metabolites. Many of them are bioactive constituents of medicinal herbs, some show antimicrobial activity (*1,2*). Many terpenes exhibit characteristic odors and tastes, e.g. in essential oils, fruits, vegetables and culinary herbs. In milk and cheese, however, terpenes have not been described among the character impact compounds. Table I lists the important aroma-active compounds of Swiss Gruyère cheese, a hard cheese which is highly appreciated in Europe. According to Rychlik and Bosset (*3,4*) the most potent odorants of Gruyère cheese, which are characterized by high odor activity values, are sulfur-containing components derived from the amino acid metabolism of the ripening cultures, such as methanethiol, methional and dimethyltrisulfide, as well as phenylacetic acid and 2- and 3-methylbutanal.

However, terpenes are useful markers to determine the origin of dairy products such as milk and cheese, i.e. to discriminate between highland and lowland produce. The terpene profile proves a useful authenticity marker for products with a protected designation of origin, so-called PDO products as described by Zeppa and co-workers (*5*), Buchin and co-workers (*6*), and very recently by Cornu and co-workers (*7*).

Mariaca and co-workers (*8*) investigated the volatile mono- and sesquiterpenoids in different highland and lowland plant species. They could show that highland pastures show a wide variety of different plant species. In particular dicotyledonous plants usually rich in terpenoids are more abundant in highland compared to lowland pastures. The latter are often composed of plant species containing less dicotyledons and thus, are poor in terpenes. Bosset and co-workers (*9*) compared hard cheese produced in the highlands to hard cheese of the lowlands and found statistically significant differences in the chemical composition such as terpenes, fatty acids and triglycerides between the highland and lowland cheeses.

Only little, however, is known about the influence of rumen fermentation on the monoterpenes present in plants and the changes they undergo from pasture to cheese. The objective of the present study was to trace the monoterpene composition of fresh grass, milk and a Raclette-type cheese produced from the milk using dynamic headspace (purge + trap, P+T) gas chromatography in combination with mass spectrometry (GC-MS).

To our knowledge this is the first study to investigate the influence of rumen fermentation on the degradation of monoterpenes present in a model plant commonly found on pastures. The results of the *in vitro* study will be compared to the volatiles found in milk and in cheese.

A chiral stationary phase was employed to study whether the enantiomeric ratios of the monoterpenes found in pasture are maintained in milk and in cheese, and to investigate the impact of the bovine rumen microflora on the terpene

Table I. Aroma compounds of a lowland Gruyère cheese

Compound	*Odor descriptor*	*Cnc µg/kg*	*Odor activity value*
Methanethiol	Sulfurous	700	11300
Methional	Boiled potato-like	99	495
Dimethyltrisulfide	Cabbage-like	136	54
Phenylacetic acid	Honey-like	7270	39
2-Methylbutanal	Malty	255	26
3-Methylbutanal	Malty	219	17
2-/3-Methylbutanoic acids	Sweaty	81000	8.1
Acetic acid	Vinegar-like, pungent	298000	5.5
Butanoic acid	Sweaty	97000	5.4
δ-Decalactone	Sweet, coconut-like	1690	4.2
Propanoic acid	Fruity, pungent	87000	2.9
Phenylacetaldehyde	Honey-like	20	1
2,3-Diethyl-5-methylpyrazine	Earthy	0.6	1
Methylpropanoic acid	Sweaty, rancid	41000	0.5
(*E*)-2-Nonenal	Green	323	0.4
2-Ethyl-3,5-dimethylpyrazine	Earthy	0.7	0.3
Hexanoic acid	Goat-like	28000	0.3

Data are from references *3* and *4*.

composition of cow parsnip (*Heracleum sphondylium* L.), which was selected as model plant.

Experimental

Materials

The fresh grass samples originated from six alps in the Canton of Vaud, Switzerland, near the villages of Les Moulins and L'Etivaz, at altitudes from 1400 to 1920 m. They were collected at the end of August 2003. Seven alps located around Les Moulins (Canton of Vaud, Switzerland, altitudes of the alps: 1400 to 1900 m) provided the cow's milk (one sample each), and one Raclette-type cheese investigated in the present study had been produced from the same milk. Cow parsnip (*Heracleum sphondylium* L.) was collected twice, in mid-August 2003 near Schwarzenburg (Canton of Berne, Switzerland, altitude around 700 m) and at the end of August 2003 in La Frêtaz near St. Croix in the Swiss Jura (Canton of Vaud, Switzerland, altitude around 1200 m). The grass, the milk and the cow parsnip samples were pooled to form composite samples. The samples were kept frozen at -20 °C before use.

Rumen fluid was collected from 3 healthy cows aged between 5 and 8 years with an artificial fistula connected to the rumen. The cows had been fed with hay *ad libitum* for two weeks prior to the study.

The composite grass sample (5 g) was roughly chopped and mixed with MilliQ water (100 mL) in a Polytron PT 3000 mixer (Kinematica, Littau, Switzerland) at 10000 rpm for 2 min. The suspension (2 mL) was diluted with MilliQ water (100 mL) and 2 mL of the diluted suspension were pipetted into a 25 mL non-fritted sparger (Schmidlin, Neuheim, Switzerland) for the P+T analysis. Cow parsnip (2 g) was ground and suspended in phosphate buffer pH 6.6 (30 mL) at the same pH as the rumen fluid according to (*10*). Rumen fluid (15 mL) was added and the suspension incubated at 39 °C for 24 h in a shaking mixer under argon atmosphere to prevent oxidation. The suspension (1 mL) was diluted with MilliQ water (100 mL) and 2 mL of the dilution were employed for the P+T analysis.

Milk (20 mL) was directly pipetted into the sparger for subsequent P+T analysis. The Raclette-type cheese was finely grated and 5 g were weighed into a beaker, mixed with MilliQ water (45 mL) and homogenized at 10000 rpm (Polytron) for 2 min. Immediately after homogenization the sample (2 g) was weighed into a sparger.

Instrumental Analysis

Dynamic headspace extraction (Purge + trap, P+T) using a LSC 2000 P+T system equipped with a cryofocusing unit and a trap no. 8 (Tekmar, Cincinnati, OH) was employed for the extraction of the monoterpenoids according to an adapted method from Mariaca and co-workers (*8*). The samples were weighed into a 25 mL non-fritted sparger (Schmidlin, Neuheim, Switzerland) and extracted at 40 °C using a water bath. The P+T conditions were as follows: nitrogen was used as purge gas at 41 mL/min purge flow (vent); purge 20 min; dry purge 11 min; cap cool down –150 °C; desorb 4 min at 240°C; inject 1.5 min at 215 °C; bake 10 min at 260 °C; transfer line to GC 150 °C.

Gas chromatography coupled to mass spectrometry (GC-MS) for the separation and analysis of the monoterpenes was conducted on an Agilent 5890 Series II instrument (Palo Alto, CA) equipped with a 5971A MSD and a chiral capillary column CP-Chirasil-Dex CB (25 m × 0.25 mm × 0.25 μm film thickness; Stehelin, Basel, Switzerland) using helium as carrier gas at 50 kPa (1.2 mL/min) at 35 °C. The temperature program ran from 35 °C (0.5 min hold) to 215 °C at a linear gradient of 2.5 °C/min. The temperature of the GC-MS transfer line was set to 280 °C, the MSD scanned in the electron impact (EI) mode at 70 eV from *m/z* 26 to 350 at 0.8 scans/s. Terpenes were identified by their identical linear GC retention indices and mass spectra with the ones of reference terpenes (Fluka, Buchs, Switzerland). The terpene metabolites were tentatively identified based upon comparison of their mass spectra with commercial mass spectrometry databases and retention indices in the literature. Single ion monitoring (SIM) of the ions *m/z* 93, 95, 119, 70, or 69 for the monoterpenes and the terpene metabolites, respectively, was employed for semiquantitation comparing their peak heights before and after incubation with rumen fluid. The ions *m/z* 136 and 93 were selected as qualifier ions for the monoterpenes and their metabolites, respectively. The samples were analyzed once. The software SPSS version 11 (SPSS Inc., Chicago, IL) was used for correlating the degradation of (*E*-)-β-ocimene, δ-3-carene and β-myrcene and the formation of menth-1-ene, 3,7-dimethyl-1,6-octadiene and 3,7-dimethyl-2-octene. Except for the correlation the data were not statistically treated.

Results and Discussion

The grass sample consisted of plants of various families as described in (6), i.e. *Apiaceae, Asteraceae, Campanulaceae, Rosaceae, Fabaceae, Geraniaceae, Lamiaceae, Plantaginaceae, Ranunculaceae, Polygonaceae, Rubiaceae*, and also *Poaceae*, the family to which common grass belongs. The term "grass" will be used for the mixture of the mentioned plant species. The monoterpene composition of fresh grass was analyzed by dynamic headspace gas chromatography-mass spectrometry using a chiral stationary phase. β-Myrcene, p-cymene, the two enantiomers (+)- and (–)-limonene, (*E*-)-β-ocimene and (*Z*)-β-ocimene, γ-terpinene and (-)-α-pinene were the predominant monoterpenes in the grass samples analyzed (data not shown). Figure 1 depicts the structures of the monoterpenes detected in the present study. Sabinene, α-thujene, (+)-α-pinene, α-phellandrene, the two enantiomers (–)-β-pinene and (+)-β-pinene, and α-terpinolene were only minor components. Table II compares the monoterpenoids found in fresh grass, in milk, and in the Raclette-type cheese.

In the milk samples collected from farms from the same sites the additional monoterpenes α-terpinene, δ-3-carene as well as α-phellandrene and (–)-camphene were detected as minor constituents, however, no (*E*)-β-ocimene was found. Interestingly, the analysis revealed further monoterpenoids tentatively identified as 3,7-dimethyl-1,6-octadiene, 2,6-dimethyl-2,6-octadiene, 3,7-dimethyl-2-octene, and camphane. These monoterpenoids had been absent in fresh grass. In the Raclette-type cheese produced from the milk basically the same monoterpenes were found as in milk, except for δ-3-carene. Additionally, 2-carene, (*E*)-β-ocimene and the tentatively identified menth-1-ene were detected. 3,7-Dimethyl-1,6-octadiene, 2,6-dimethyl-2,6-octadiene, 3,7-dimethyl-2-octene, camphane and menth-1-ene were only detected in the milk and cheese

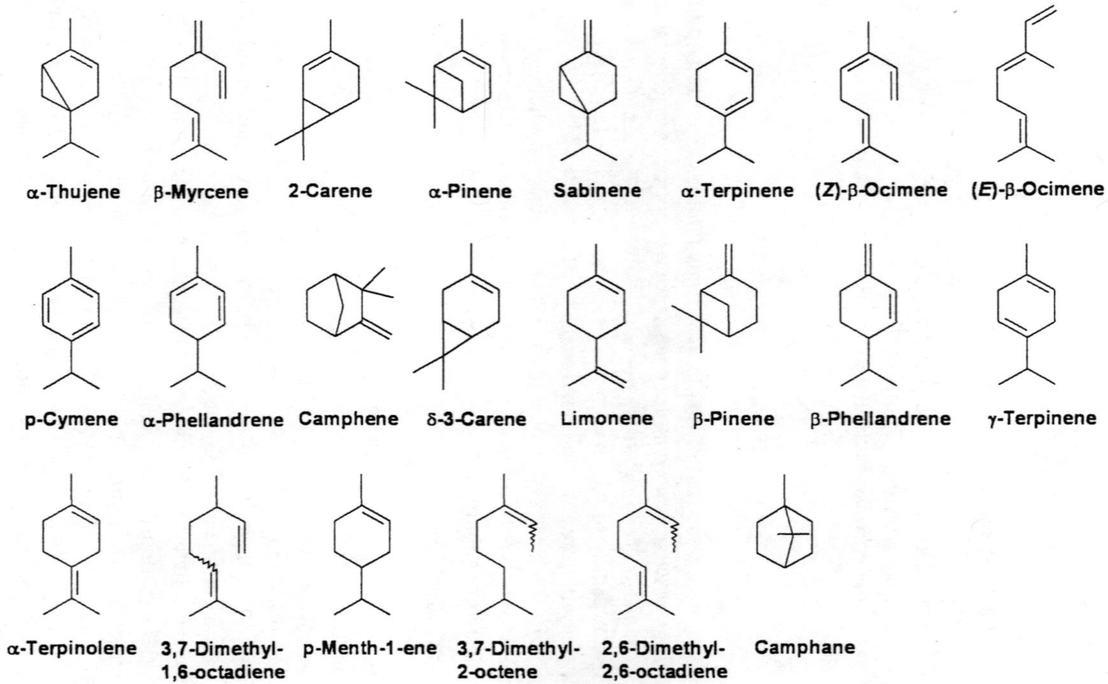

Figure 1. Structures of monoterpenes and monoterpene metabolites tentatively identified in grass, milk, Raclette-type cheese and in cow parsnip

Table II. Monoterpenoids detected in grass, milk and Raclette-type cheese

Compound	*Grass*	*Milk*	*Raclette-type cheese*
α-Thujene	✓	✓	✓
β-Myrcene	✓	✓	✓
2-Carene	-	-	✓
(-)-α-Pinene	✓	✓	✓
Sabinene	✓	✓	✓
(+)-α-Pinene	✓	✓	✓
α-Terpinene	-	✓	✓
(*Z*)-β-Ocimene	✓	✓	✓
(+)-β-Pinene	✓	✓	✓
(*E*)-β-Ocimene	✓	-	✓
δ-3-Carene	-	✓	-
p-Cymene	✓	✓	✓
α-Phellandrene	✓	✓	✓
(-)-Camphene	✓	✓	✓
(+)-Camphene	-	✓	✓
(-)-Limonene	✓	✓	✓
(+)-Limonene	✓	✓	✓
(-)-β-Pinene	✓	✓	✓
β-Phellandrene	-	✓	✓
γ-Terpinene	✓	✓	✓
α-Terpinolene	✓	✓	✓
3,7-Dimethyl-1,6-octadiene	-	✓	✓
Menth-1-ene	-	-	✓
3,7-Dimethyl-2-octene	-	✓	✓
2,6-Dimethyl-2,6-octadiene	-	✓	✓
Camphane	-	✓	✓

samples, however not in fresh grass. The presence of these compounds had already been described in alpine milk by Buchin and co-workers (*6*).

The enantiomeric ratios of selected monoterpenes were analyzed to examine whether the ratios changed from pasture to cheese, but no consistent trend could be observed. In the grass sample (−)-α-pinene was prevalent over the (+)-enantiomer, and (+)-limonene over (−)-limonene. In contrast, in milk and in cheese (+)-α-pinene predominated. As already found in the grass sample, (+)-limonene prevailed also in milk and in the cheese samples over (−)-limonene. Presumably, the cows had grazed on a pasture with plants producing more (+)-α-pinene.

Incubation of cow parsnip with rumen fluid

The exclusive presence of the dimethyl octadienes, menth-1-ene and the other tentatively identified compounds in milk and cheese raised the hypothesis that these compounds might have been formed during rumen fermentation. Many of the monoterpenes shown in Figure 1 show two or three double bonds, whereas 3,7-dimethyl-1,6-octadiene, 2,6-dimethyl-2,6-octadiene, 3,7-dimethyl-2-octene, camphane and menth-1-ene have two or only one double bond. Hydrogenation during rumen fermentation might be involved in their formation. The rumen flora is known to be strictly anaerobic. It produces hydrogen among other gases and biohydrogenation reactions convert e.g. unsaturated fatty acids into saturated fatty acids (*11*). Cow parsnip, *Heracleum sphondylium* L., was selected as a model to investigate a potential correlation between the degradation of certain monoterpenes and the formation of the above-mentioned compounds. Cow parsnip belongs to the family *Apiaceae*, is found abundantly in pastures and rich in terpenoids as described by Mariaca and co-workers (*8*).

Ground cow parsnip was incubated with rumen fluid over 24 h and monitor the changes in its monoterpene composition. After 12 h of fermentation all monoterpenes were at least partially degraded. The rate and extent of degradation varied for the different compounds. Figure 2 illustrates the degradation of selected monoterpenes after 12 h of incubation.

(*Z*)-β-ocimene, (*E*)-β-ocimene, α-phellandrene and β-myrcene were completely or almost completely degraded, while around 40 % of (+)-limonene and p-cymene, and 20 to 35 % of most other monoterpenes remained after 12 h of incubation.

During incubation with rumen fluid the formation of the monoterpenoid compounds that had been detected in milk and in cheese was observed. The extent and rate of formation of these compounds varied, as it could be seen for the degradation of the different monoterpenes.

Statistically significant ($p < 0.05$) negative correlations were found for the degradation of (*E*)-β-ocimene and the co-eluting δ-3-carene and the β-myrcene and the formation of menth-1-ene, 3,7-dimethyl-1,6-octadiene and 3,7-dimethyl-2-octene. (*E*)-β-Ocimene, δ-3-carene and β-myrcene were almost completely and rapidly degraded during the first 8 h of incubation. Menth-1-ene increased steadily during the first 8 h of incubation, then was slowly degraded from 8 to 12 h of rumen fermentation. After 12 h the menth-1-ene signal intensity dropped rapidly over the next 4 h as shown in Figure 3.

The formation of 3,7-dimethyl-1,6-octadiene began after 2 h of rumen fermentation and reached its maximum after 8 h. Then it degraded slightly. At first in parallel to 3,7-dimethyl-1,6-octadiene, but after 4 h of incubation only, the onset of 3,7-dimethyl-2-octene could be observed. The peak signal increased rapidly after 12 h of rumen fermentation.

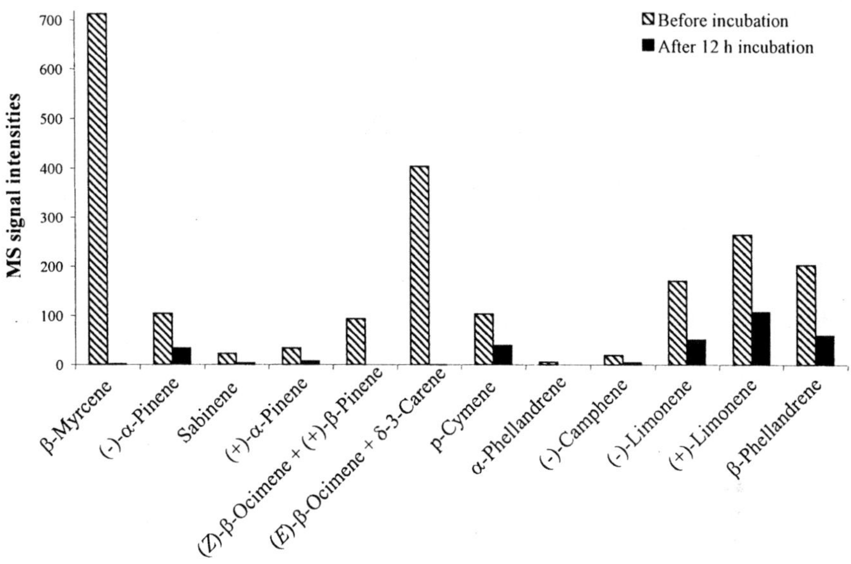

Figure 2. Monoterpenoids of cow parsnip before and after 12 h of incubation with rumen fluid

The formation of the newly formed compounds may be explained by hydrogenation reactions by the rumen flora or by enzymes present in the rumen fluid as schematically presented in Figure 2. Hydrogenation of the double bond at position 3 of β-myrcene or of (*Z*)- and (*E*)-β-ocimene yields 3,7-dimethyl-1,6-octadiene, a shift of the double bond at position 1 might explain the formation of 2,6-dimethyl-2,6-octadiene which had also been detected in the milk and Raclette-type cheese samples. A further hydrogenation of 2,6-dimethyl-2,6-octadiene yields 3,7-dimethyl-2-octene.

Conclusions

The monoterpene composition of fresh grass, milk, and cheese was analyzed by dynamic headspace GC-MS. It was shown that milk and cheese revealed additional monoterpenes which had not been detected in the grass samples. A model plant, cow parsnip, *Heracleum sphondylium* L., was incubated i*n vitro* with rumen fluid to investigate for the first time the relationship between the degradation of monoterpenes and the formation of these additional monoterpenoids. The monoterpenes of cow parsnip were at least partially degraded during incubation with rumen fluid and 3,7-dimethyl-1,6-octadiene, 3,7-dimethyl-2-octene, 2,6-dimethyl-2,6-octadiene together with camphane and

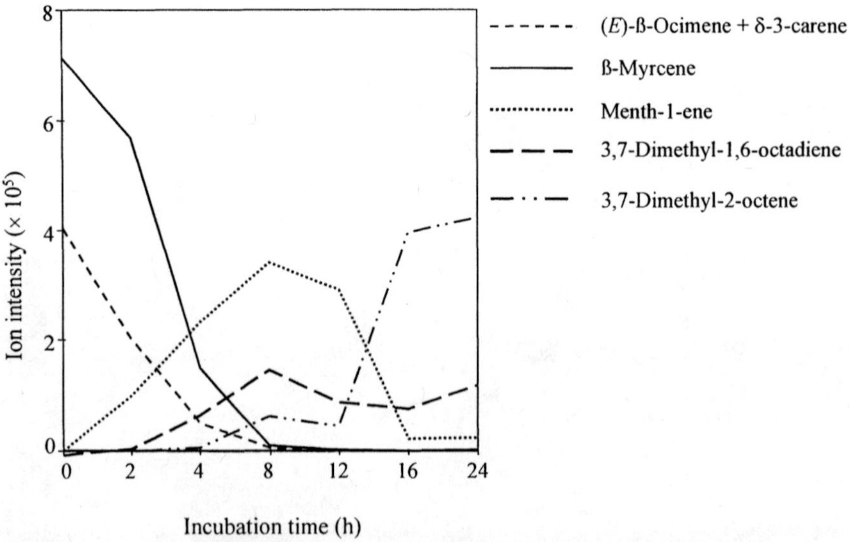

Figure 3. Degradation and new formation of selected monoterpenoids during incubation of Heracleum sphondylium L.with rumen fluid.

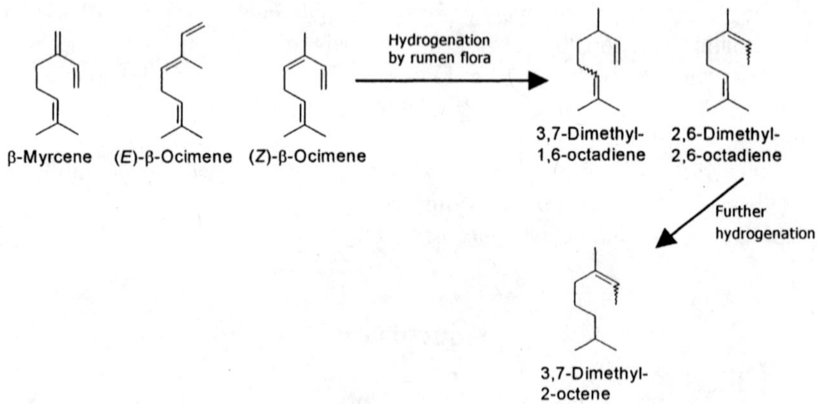

Figure 4. Hypothetic schematic formation of selected monoterpenoids by hydrogenation

menth-1-ene were formed. It is hypothesized that β-myrcene, (*E*)- and (*Z*)-β-ocimene are hydrogenated during incubation with rumen fluid to yield these components. The results show that monoterpenes are partially degraded during rumen fermentation and new compounds are formed, which are also transferred into milk and cheese.

Acknowledgments

The authors thank Roland Gauch and Bernard Jeangros for their help collecting the samples, Kurt Zbinden for providing the rumen fluid and Frigga Dohme for useful discussions.

References

1. Nostro, A.; Germanò, M.P.; D'Angelo, V.; Marino, A.; Cannatelli, M.A. *Lett. Appl. Microbiol.* **2000**, *30*, 379-384.
2. Delgado, B.; Palop, A.; Fernández, P.S.; Periamo, P.M. *Eur. Food Res. Technol.* **2004**, *218*, 188-193.
3. Rychlik, M.; Bosset, J. O. *Int. Dairy J.* **2001**, *11*, 895-901.
4. Rychlik, M.; Bosset, J.O. *Int. Dairy J.* **2001**, *11*, 903-910.
5. Zeppa, G.; Giordano, M. In Abstracts of the 5th International Meeting on Mountain Cheese, INRA, Ed., Arèches, France, 2003
6. Buchin, S.; Salmon, J.C.; Carnat, A.P.; Berger, T.; Bugaud, C.; Bosset, J.O. *Mitt. Lebensm. Hyg.* **2002**, *93*, 199-216.
7. Cornu, A.; Kondjoyan, N.; Martin, B.; Verdier-Metz, I.; Pradel, P.; Berdagué, J.L.; Coulon, J.B. *J. Sci. Food Agric.* **2005**, *85*, 2040-2046.
8. Mariaca R.; Berger T.; Gauch R.; Imhof M., Jeangros B.; Bosset J.O. *J. Agric. Food Chem.* **1997**, *45*, 4423-4434.
9. Bosset J.O.; Berger T.; Bütikofer U.; Collomb M.; Gauch R.; Lavanchy P.; Sieber R. *Agrarforschung* **1998**, *5*, 363-366.
10. Fievez, V.; Dohme F.; Danneels, M.; Raes K.; Demeyer, D. *Anim. Feed Sci. Technol.* **2003**, *104*, 41-58.
11. Collomb M.; Sieber R.; Bütikofer U. *Lipids* **2004**, *39*, 355-364.

Chapter 8

Volatile Sulfur Compounds in Cheddar Cheese Determined by Headspace Solid-Phase Microextraction–Gas Chromatography–PFPD

Michael C. Qian and Helen M. Burbank

Department of Food Science and Technology, Oregon State University, 100 Wiegand Hall, Corvallis, OR 97331

A sensitive headspace SPME-GC-PFPD method has been developed for volatile sulfur analysis in cheese. A Carboxen-PDMS solid-phase microextraction (SPME) fiber (85μm) was employed to extract volatile sulfur compounds from the cheese matrix and the volatile sulfur compound were analyzed by gas chromatography-pulsed flame photometric detection. The highly reactive thiols were successfully stabilized. An evaluation of fiber exposure time, temperature of extraction, and sample size was undertaken in order to determine effective extraction conditions. Calibration curves were built with spiked cheese samples and quantification of volatile sulfur compounds was achieved by using multiple internal standards. Commercial cheese samples manufactured from different regions of US with different aging were studied. Although the sulfur profile varied considerably among the cheese brands, the concentrations of certain sulfur compounds, namely hydrogen sulfide, methanethiol, and dimethyl trisulfide, were found to directly correlate with age.

Importance of Volatile Sulfur Compounds to Cheddar Cheese Aroma

Volatile sulfur compounds (VSCs), most with low odor thresholds, are considered important flavor contributors to many varieties of cheeses (*1-5*). The presence of VSCs in cheese is mainly the result of the biodegradation of sulfur-containing amino acids. The formation of VSCs by bacteria, such as *Lactococcus lactis* subsp. *cremoris* and *Brevibacterium linens*, has been thoroughly investigated (*6*), however, the exact mechanisms that lead to these flavor compounds are still debated by many researchers. The ripening process of cheese involves, in part, the decomposition of sulfur-containing amino acids, cysteine and methionine. An increase in the concentration of methanethiol as Cheddar cheese ages has been reported (*7*) and it is postulated that other sulfur-containing compounds may also follow a similar trend with aging.

Methanethiol is thought to be produced via a single- or two-step pathway during the catabolism of L-methionine. The single-step route is thought to be enzymatically catalyzed by L-methionine γ-lyase (*8*) or cystathionine β-lyase activity (*9*). The two-step pathway could be caused by the transamination of L-methionine when in the presence of α-keto glutarate (*10-12*). This results in the formation of α-keto γ-methyl thiobutyrate (KMTB) which can then be further broken down enzymatically to form methional and methanethiol; however, the exact enzymes involved in this second step are still unclear. It has also been reported that chemical decomposition of methionine can produce methanethiol due to the low oxidation-reduction potential of cheese (*13*). Methanethiol not only contributes to Cheddar cheese flavor but also is a precursor for several other sulfur compounds.

Hydrogen sulfide, methanethiol, dimethyl sulfide, dimethyl disulfide, dimethyl trisulfuide and methional have all been found to be significant to the flavor of Cheddar cheese (*14-18*). Many of these VSCs are described to have odors of strong garlic, decomposing cabbage or decomposing protein when smelled alone. However, when they are mixed with other volatile compounds, their unpleasant characteristic odors often disappear and can actually contribute to agreeable flavor notes.

While there are reports suggesting that hydrogen sulfide is an important flavor contributor to Cheddar cheese (*19*), others suggest that rather than acting as a flavor contributor, the presence of hydrogen sulfide is merely an indicator that some important milk enzymes are still active (*20*). Detection and measurement of hydrogen sulfide is very difficult due to its extremely low boiling point (-60°C). Traditional aroma extraction and concentration techniques will often discriminate against this compound. Consequently, accurate quantification of hydrogen sulfide has been a technical challenge.

Methanethiol is considered to be a very important flavor compound in many types of cheeses and has been reported to be responsible for the typical odor of Cheddar cheese. While it is an important flavor contributor, it is also a precursor of many other important sulfur compounds such as dimethyl disulfide and dimethyl trisulfide (*21-23*). Methanethiol has a very low detection threshold (0.06 parts per billion) in lipid media and has a decomposing cabbage or decomposing protein odor. There are often difficulties with studying this compound since methanethiol can easily oxidize during analysis to form dimethyl disulfide and dimethyl trisulfide (*24*, *25*). The aroma contribution of dimethyl disulfide can be of cauliflower and onion-like odors while dimethyl trisulfide provides garlic and/or very ripe cheese notes. Dimethyl disulfide can be formed as a Strecker degradation product of methionine (*21, 22*) or through the auto-oxidation of methanethiol (*26*).

Methional, or 3-(methylthio)propionaldehyde, has also been found to be a key aroma compound of many cheeses (*4, 16, 17, 27*). Methional can be generated by Strecker degradation of methionine and can be oxidized to its acid form, from which corresponding esters can be produced. For example, methyl thioacetate has been identified in the volatiles of limburger cheese contributing to a cooked cauliflower aroma (23), methyl thiopropanoate was found to have a Camembert-like cheesy note (28), and ethyl 3-methylthiopropanoate has been found as a trace component in Parmesan cheese (*29*). Methional can also be reduced to the corresponding alcohol; 3-(methylthio)propanol, or methionol, has also been found in premium quality Cheddar cheese (*30*) and Camembert cheese (*31*). It has been reported to have a potato-like odor (*32*).

Volatile Sulfur Analysis by HS-SPME-GC-PFPD

While there is definite evidence that various sulfur compounds could be very important to Cheddar cheese aroma, their significances to Cheddar cheese flavor are still poorly understood. This is contributed to by the fact that the volatile sulfur profile of Cheddar cheese highly relies on the specific method used for extraction and/or concentration techniques. Solvent extraction has long been used as a way to effectively extract aroma compounds from complex food matrices; however, this technique will often lose all highly volatile compounds during the concentration step. Since hydrogen sulfide and methanethiol have very low boiling points, these compounds will evaporate together with the solvent and therefore cannot be properly analyzed. Static headspace technique directly takes the volatiles from the headspace of the sample and does not involve the use of organic solvents, however, this method is usually not sensitive enough to provide successful detection of sulfur compounds due to limited sample sizes (*33-35*). Dynamic headspace also directly samples the volatiles by purging the sample with a stream of inert gas where the volatiles are first collected on an adsorbent material and then are desorbed onto a GC column for analysis. However, purge-and-trap techniques will often induce artifact formation.

In recent years, headspace solid-phase microextraction (SPME) has become widely known as a successful method for the extraction of volatile aroma compounds from a variety of food matrices. SPME has high sensitivity and minimal artifact formation. With the use of SPME fibers, there is no requirement for organic solvents, and sample preparation can be completed in minimal time. While there are a growing number of available fiber coatings, the two-phase Carboxen-polydimethylsiloxane (CAR-PDMS) fiber has repeatedly demonstrated its exceptional ability to extract sulfur compounds including methanethiol and dimethyl sulfide from food (*36-42*). The non-polar PDMS phase retains analytes through adsorptive interactions while the Carboxen phase, comprised of porous synthetic carbon, effectively traps small sulfur compounds within its macro- and micropores. The analytes are then released from the fiber by thermal desorption directly onto a GC column.

In order to achieve the highest sensitivity possible for analyzing trace levels of sulfur analytes, the method of detection must also be sufficiently sensitive. Pulsed flame photometric detection (PFPD) can be used as a sulfur-specific method of detection. Flame chemiluminescence is generated by way of the pulsed flame, which is ignited, propagated and self-terminated 3 to 4 times per second. The emission profile for specific elements is both wavelength and emission time dependent. The sulfur chemiluminescence can be detected by selecting the sulfur wavelength. In addition, due to the fact that hydrocarbons complete emission early while sulfur emissions will begin at a relatively later time after combustion, the use of a 'gate delay' allows for only a particular portion of the emission signal to be processed during each pulse of the ignition-emission cycle, thus greatly reducing the flame background due to hydrocarbon emission. As a result, the sensitivity is improved and the chromatogram is much cleaner than that from traditional flame photometric detection (FPD) by displaying only the response due to sulfur-containing compounds. The ability of the PFPD to selectively process the signal improves the overall sensitivity, allowing the PFPD to detect sulfur-containing compounds at a much lower detection limit than nearly all other methods of detection (*43*). The combination of SPME with GC-PFPD greatly enhances the ability to successfully extract and detect low concentrations of VSCs in cheese.

VSCs were investigated using SPME, where a Stableflex 85 µm CAR-PDMS fiber (Supelco, Bellefonte, PA) was used in this study and prior to use, the fiber was conditioned at 300°C for 90 min. A CombiPAL autosampler (CTC

Analytics AG, Zwingen, Switzerland) fitted with a SPME adapter and vial heater/agitator along with a CP-3800 gas chromatograph (Varian, Walnut Creek, California) equipped with PFPD were used. Samples were placed into 20mL headspace vials and agitation during sample equilibration was set to 500 rpm while agitation during extraction was held at 250 rpm. Varying extraction temperatures and lengths of fiber exposure time were tested to find suitable extraction parameters.

The volatile compounds extracted by the SPME fiber were thermally desorbed in the 300°C injector port for 7 min where splitless mode was used for the first four minutes. Separation of the analytes was performed using a DB-FFAP fused silica capillary column (30 m x 0.32 mm, 1.0 μm film) with nitrogen flow at 2.0 ml/min. The oven temperature program was as follows: 35°C held for 5 min, heated to 150°C at a rate of 10°C/min, held for 1 min, then heated to 220°C at a rate of 20°C/min with a final hold time of 5 min. The PFPD was held at 300°C and 500 V with the following flow rates: H_2 at 14 mL/min, air1 at 17 mL/min, and air 2 at 10 mL/min and the detector response signals were collected and integrated using computer software.

Sulfur Standard Preparation and Stabilization

Chromatographic identification of target sulfur compounds was performed by comparing retention times to those of pure standards, most of which were obtained commercially with purity verified by observance of a single peak by GC-PFPD. Standards included dimethyl sulfide, carbon disulfide, dimethyl disulfide, dimethyl trisulfide, and methional. A standard solution of hydrogen sulfide was freshly prepared by dissolving sodium sulfide salt in distilled water (pH 10), where H_2S *(gas)* was released by adding a volume of the solution to an acidified matrix. Carbonyl sulfide (COS) was prepared by a published method (*44*), with some modification: concentrated sulfuric acid was added dropwise from a dropping funnel to potassium thiocyanate in a stoppered Erlenmeyer flask, then the generated COS *(gas)* was passed through a small diameter glass transfer-line immersed in a cold water bath (for exothermic nature of reaction) and the gas was trapped by bubbling into a separate flask containing distilled water. Standard solutions for a majority of the sulfur-containing compounds commonly found in Cheddar cheese were relatively easy to prepare, however methanethiol, a key aroma compound for Cheddar, presented much greater difficulties during analysis.

Several commercially obtained methanethiol standards, including mercaptan sodium salt and methanethiol in methanol, were obtained and tested for purity by

SPME-GC-PFPD. Unfortunately, it was quickly discovered that these standards were far from pure since large amounts of dimethyl disulfide, and to a lesser extent dimethyl trisulfide, were observed in the chromatograms. This was evidence that oxidation had already occurred during commercial preparation and/or storage of the standards. Further experiments demonstrated that methanethiol standard solutions could be freshly prepared from pure gas.

In order to investigate the stability of methanethiol, various solvents were tested during the preparation of the standard, including water, methanol, butanol, a phosphate buffer and an organic acid solution (both at pH 5.15), and vegetable oil. The methanethiol gas was bubbled into each solvent separately and appropriate dilutions were made to achieve low ppb concentrations. To determine the purity/stability of each, a small aliquot of each diluted standard solution was placed into an autosampler vial and analyzed by SPME-GC-PFPD using two different extraction temperatures, 30°C and 50°C. The resulting chromatograms are shown in Figure 1.

When water was used as a solvent, it was found that methanethiol quickly oxidized to dimethyl disulfide, especially when the higher temperature was used during SPME extraction. In fact, with extraction at 50°C, it appeared that the presence of water induced the complete transformation of the thiol to the disulfide (Figure 1, A1) in contrast to the smaller amount of disulfide formed at 30°C (Figure 1, A2). Although cheese is comprised of a large aqueous portion, its composition also contains a substantial amount of lipids. Therefore, the stability of methanethiol in vegetable oil was tested. From the resulting chromatogram, it appeared that methanethiol in vegetable oil was stable, where negligible amounts of disulfides were present (Figure 1, B). However, vegetable oil was determined to be an unacceptable solvent for VSCs due to its high viscosity.

Some alcohols with differing chain lengths were then tested. A methanethiol solution freshly prepared in methanol was found to have good stability where minimal amount of disulfide was formed at either extraction temperature (Figure 1, C). Butanol was also examined for methanethiol stability however, instead of observing the formation of dimethyl disulfide, it was discovered that carbon disulfide was present after SPME extraction (Figure 1, D). Analysis by mass spectrometry confirmed the identity of carbon disulfide; however, the cause for this oxidation reaction in the presence of butanol is currently unclear.

The effects of antioxidants on the stability of methanethiol were investigated using aqueous buffer solutions. Phosphate buffer at pH 5.1, similar to the natural pH of cheese, did not possess acceptable stabilizing characteristics for methanethiol since a sizeable amount of dimethyl disulfide was formed during extraction (Figure 1, E). Nevertheless, the presence of the phosphate ions did appear to play some roles in stabilizing methanethiol since extraction at 50°C did not result in complete oxidation as seen with plain water. Similarly, the organic

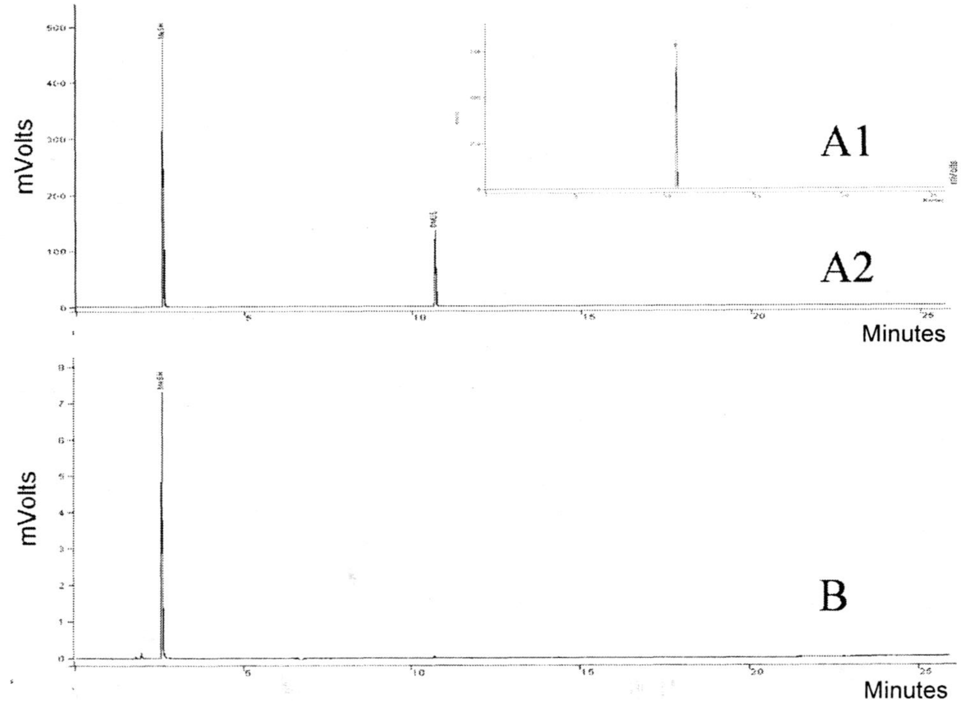

Figure 1. Chromatograms obtained using SPME-GC-PFPD for methanethiol solutions prepared with different solvents, as follows: A) water, SPME extraction at 50°C (A1) and at 30°C (A2); B) vegetable oil; C) methanol; D) butanol; E) phosphate buffer at pH 5.1; F) organic acid buffer at pH 5.15. [Note: if no temperature specified, results were the same for both tested temperatures (30°C and 50°C).] *Continued on next page.*

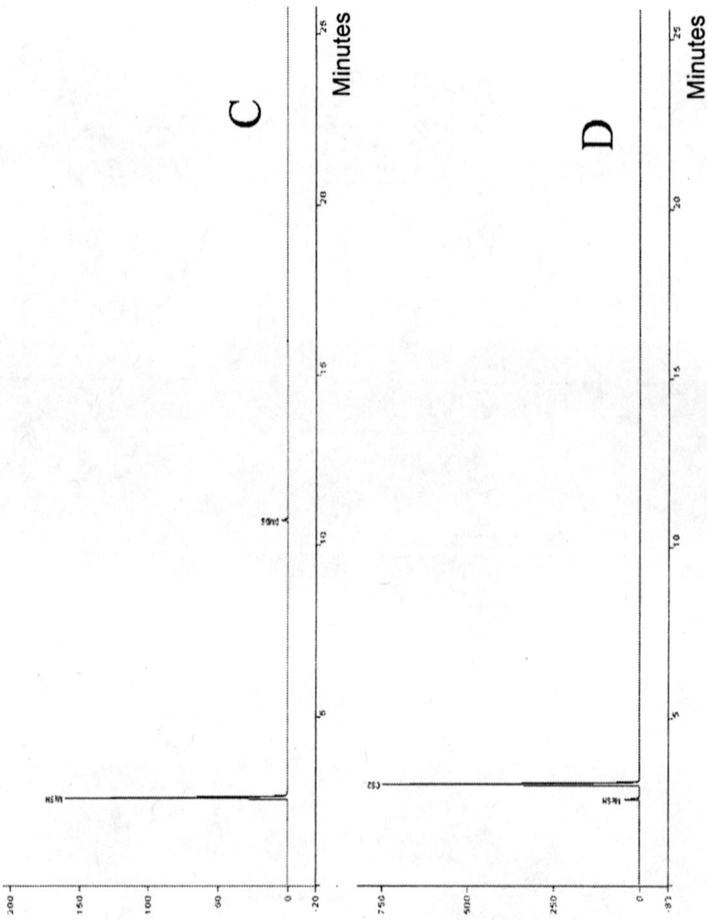
C
D
mVolts
mVolts
Minutes
Minutes

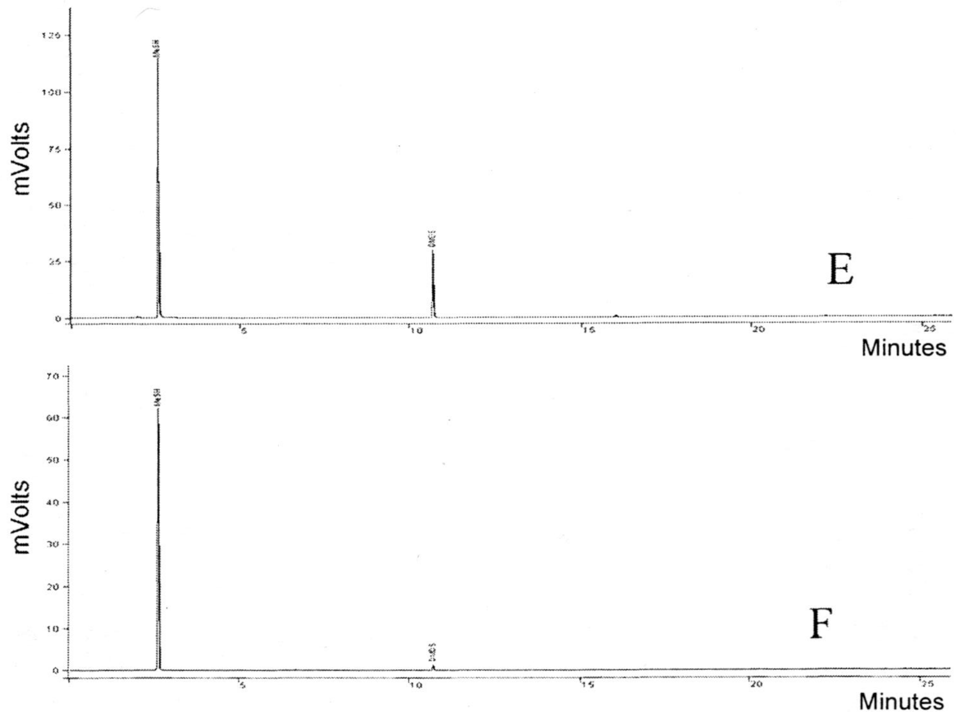

Figure 1. Continued.

acid solution (aqueous tartaric acid adjusted to pH 5.1 with dilute NaOH) also showed some success in preventing oxidation where a reduced amount of dimethyl disulfide formation was observed (Figure 1, F).

It was concluded that a standard solution of methanethiol would best be prepared in methanol due to its effective solubility in this solvent. However, even with its relative stability in methanol, methanethiol can still become oxidized to dimethyl disulfide after short-term storage. Therefore, all methanethiol standard solutions were freshly prepared weekly or as needed. Further experiments discovered that a combination of low pH, addition of tartaric acid, derivatization of the GC injection liner and exclusion of oxygen with inert gas during analysis could effectively stabilize methanethiol and allow for its accurate determination in cheese (*24*).

Standard Curve preparation of Volatile Sulfur Compounds

For the analysis of the cheese samples, SPME extraction parameters were evaluated in order to obtain satisfactory sensitivity. Extraction time, temperature, and sample size were all taken into consideration for overall extraction efficiency (*24*). Extraction time study demonstrated that as time increased so did the amount of volatiles extracted by the SPME fiber, but the relationship was not the same for all sulfur compounds. While most sulfur compounds seemed to follow a logarithmic-type trend, the data for dimethyl disulfide and dimethyl trisulfide was better fit with a power function. In all cases, adsorption rates progressively decreased with no immediate evidence of reaching equilibrium within the tested time range (up to 2 h) (Figure 2). For many VSCs, significant improvement for extraction efficiency was observed when the fiber exposure time was increased from 5 min to 30 min. The extraction efficiency improved less significantly with further increases in exposure time. Although increased fiber exposure time generally resulted in larger peak areas, lengthy extraction was found to be unnecessary due to the enhanced sensitivity of the PFPD detector. For practical reasons, an extraction time of 30 min was chosen as a good compromise between overall sensitivity and runtime efficiency for the instrument.

A relationship between temperature and extraction efficiency was not easy to establish for all sulfur compounds. Higher temperatures typically increase the volatility of volatile compounds, thus increasing their concentrations in the headspace. A definite increase in response occurred for dimethyl sulfone and methional when higher temperatures were used (Figure 3a). However, for some highly volatile compounds, further increase in temperature is expected to have

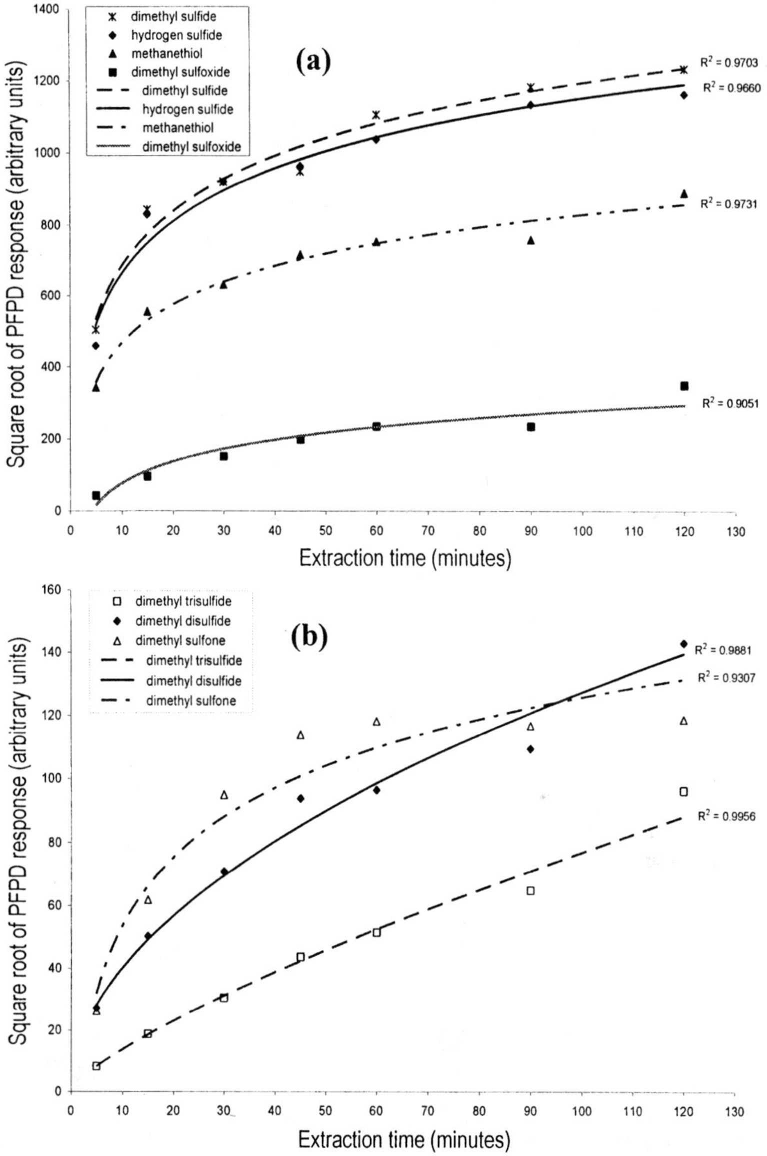

Figure 2. Effect of time on SPME extraction of VSCs from Cheddar cheese with temperature held at 50°C: (a) hydrogen sulfide, methanethiol, dimethyl sulfide, dimethyl sulfoxide; (b) dimethyl disulfide, dimethyl trisulfide, dimethyl sulfone. (Reprinted from Reference 25 with permission. Copyright 2005, Elsevier)

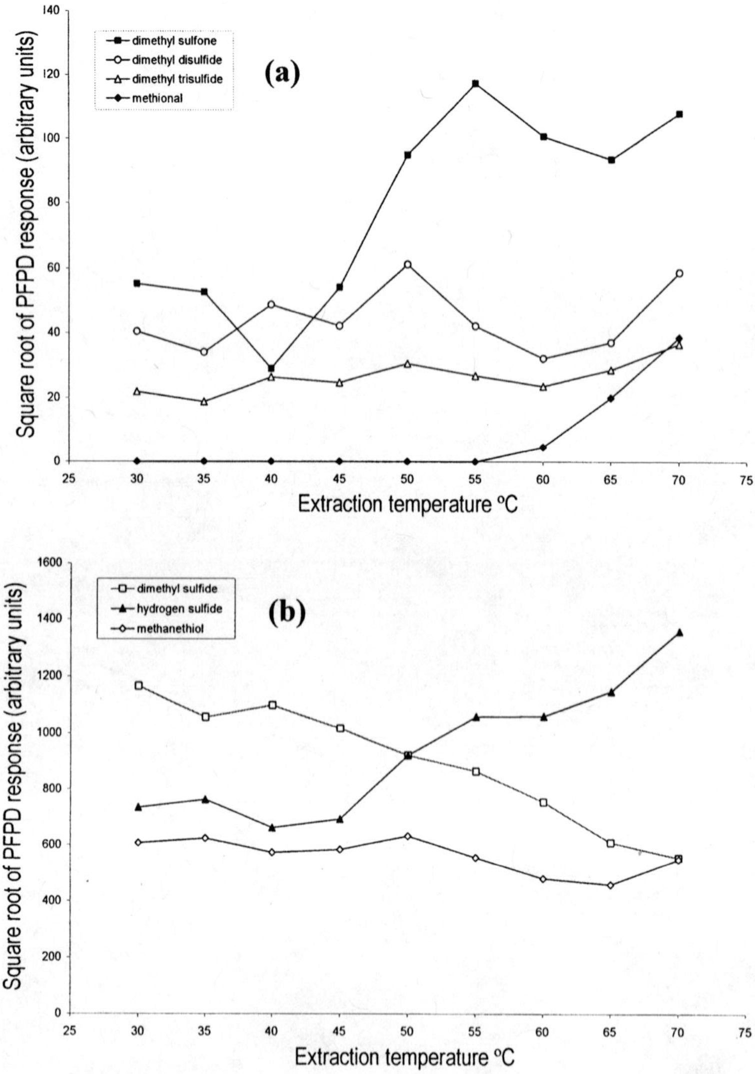

Figure 3. Effect of temperature on SPME extraction of VSCs from Cheddar cheese with extraction time set at 30 min: (a) dimethyl disulfide, dimethyl trisulfide, methional, dimethyl sulfone (b) hydrogen sulfide, methanethiol, dimethyl sulfide (Reprinted from Reference 25 with permission. Copyright 2005, Elsevier)

minimal impact on their concentrations in the headspace. This hypothesis was confirmed for methanethiol, dimethyl disulfide and dimethyl trisulfide. As temperature increased from 30°C to 70°C, the responses for these compounds stayed almost the same (Figure 3b). Uniquely, dimethyl sulfide was observed to have a continuous decrease in response with increasing temperature. Although the actual mechanism is still not clear, it has been reported that high temperature can have undesirable effects on the extraction efficiency of SPME fibers for some low molecular weight analytes with higher volatility (*33, 45, 46*). This could be related to the exothermic nature of adsorption of analytes onto the fiber, where higher temperatures may actually decrease the adsorption (*38, 46*). In the end, a satisfactory extraction temperature of 50°C was selected, where most sulfur compounds of interest gave relatively high responses with low coefficient-of-variation (CV) values.

In addition to time, temperature, and distribution coefficients, the total amount of analytes that can be adsorbed by the fiber can be directly related to the sample size (*47*). As illustrated in Figure 4, the sensitivity improved as the sample size increased from 0.5 g to 2 g. Beyond that, further increase in sample size did not significantly increase overall sensitivity. For some compounds, such as dimethyl sulfide, dimethyl disulfide and dimethyl trisulfide, increasing the sample size beyond 2 g actually decreased sensitivity. Since agitation is not easy with a solid-type matrix, large variations were observed with 5 g sample size since a uniform release of volatiles from larger sample sizes is more difficult.

Quantification was performed by a standard addition method where calibration curves were created by spiking cheese samples with a range of known concentrations of hydrogen sulfide, methanethiol, dimethyl sulfide, dimethyl disulfide, and dimethyl trisulfide. Two internal standards were used to build the calibration curves: ethyl methyl sulfide for the first three VSCs and methyl propyl disulfide for the latter two VSCs. One gram of cheese, with the addition of the internal standards along with 2mL of organic acid buffer solution, was pre-equilibrated at 50°C for 10 min and extracted at the same temperature for 30 min. Ratios of the standard to its corresponding internal standard were plotted where the y-axis represented the peak area ratio (square root taken) and the x-axis represented the concentration ratio. A chromatogram of the sulfur profile for Cheddar cheese obtained by SPME-GC-PFPD is displayed in Figure 5.

Volatile Sulfur Analysis in Cheddar Cheese

Twenty five commercially obtained Cheddar cheeses of varying age were analyzed. These commercial Cheddar cheeses could be considered as

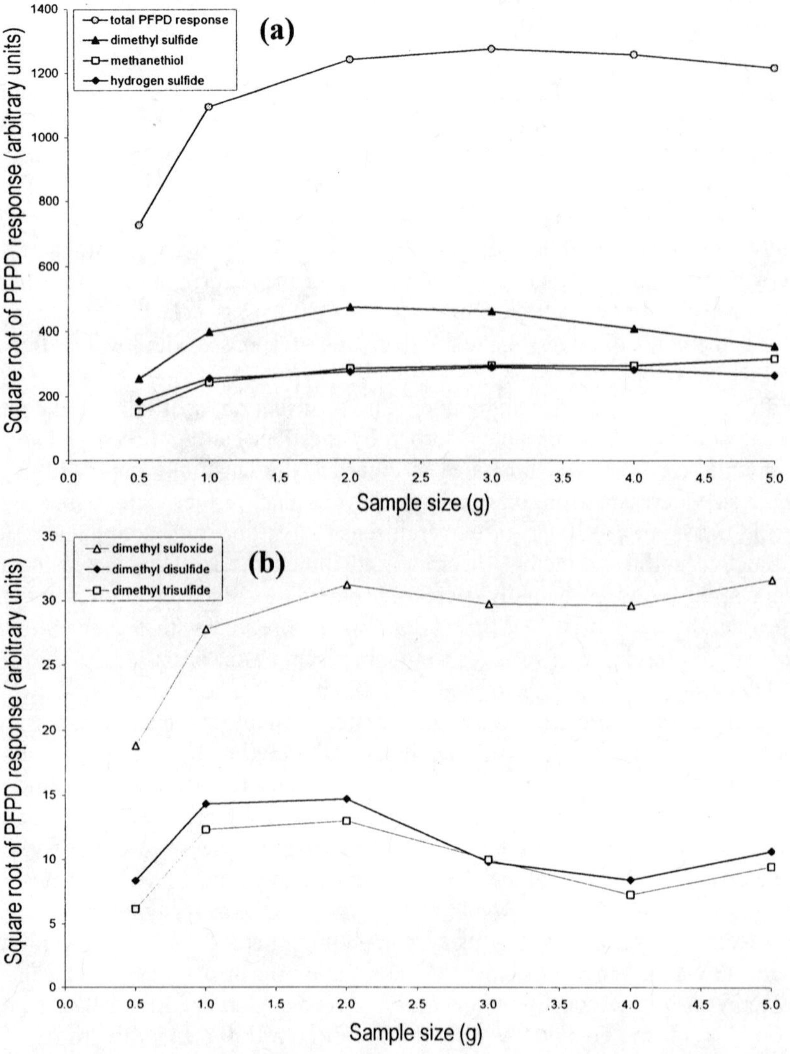

Figure 4. Effect of sample size on SPME extraction of VSCs from Cheddar cheese, extraction performed at 50°C for 30 min: (a) hydrogen sulfide, methanethiol, dimethyl sulfide, and total PFPD response (peak areas of all VSCs of interest combined, then square rooted); (b) dimethyl disulfide, dimethyl trisulfide, dimethyl sulfoxide (Reprinted from Reference 25 with permission. Copyright 2005, Elsevier)

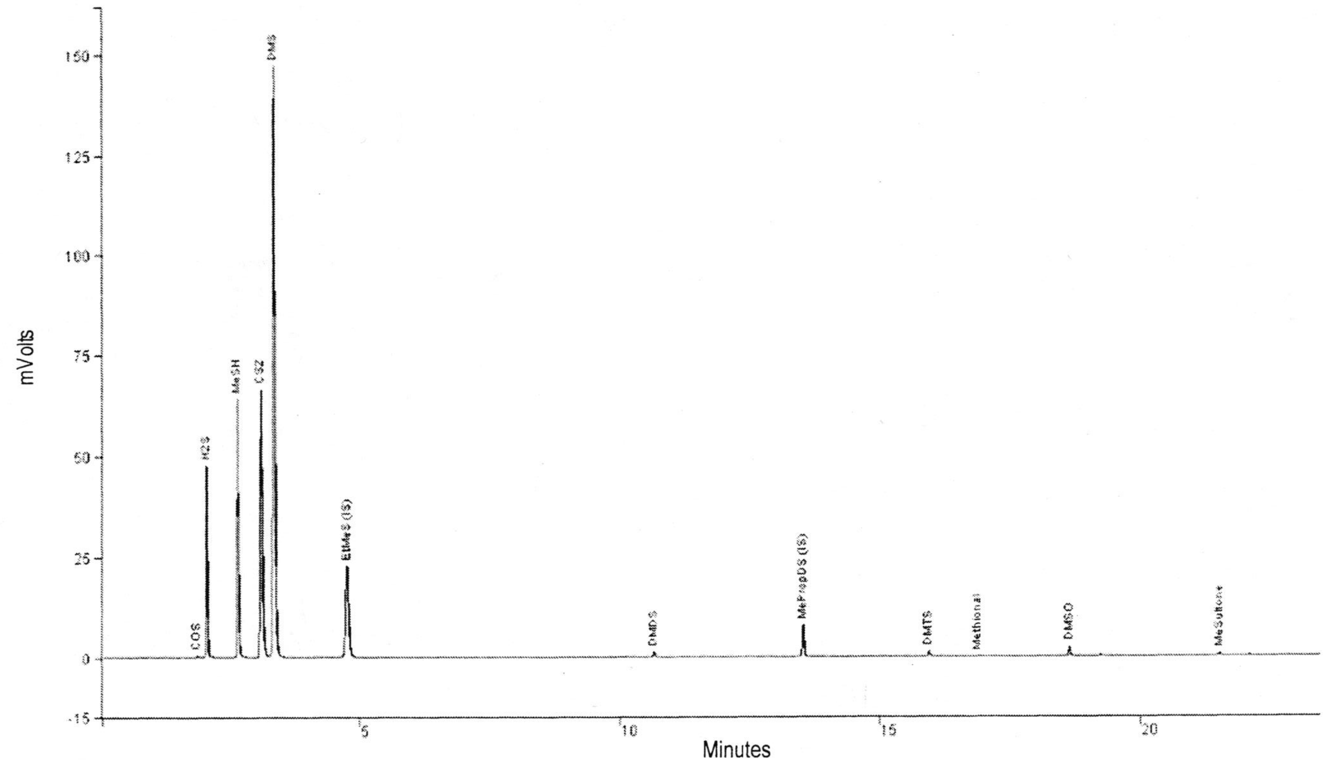

Figure 5. Chromatogram of sulfur profile for Cheddar cheese obtained by SPME-GC-PFPD.

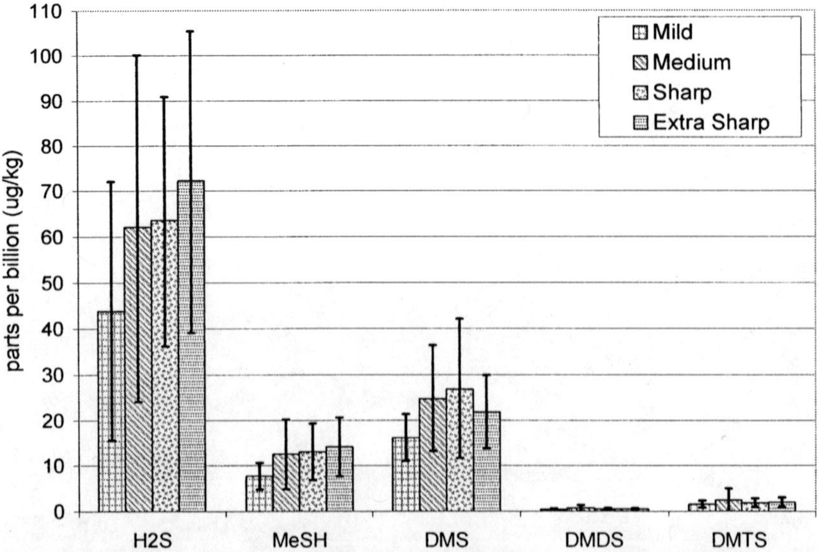

Figure 6. Average concentrations of VSCs in different ages of commercially obtained Cheddar cheese. Error bar represents range of each VSC found for that particular labeled age.

representative of what is available for consumer purchase. Therefore, the average sulfur concentrations determined for each labeled age of cheese ['mild' (n=5); 'medium' (n=5); 'sharp' (n=10); and 'extra sharp' (n=5)] provide a fairly accurate view of the overall sulfur content that can be expected in Cheddar cheese. Of the selected cheeses analyzed thus far, definite brand-to-brand distinctions exist. This lends to the fact that each manufacturer uses particular milk sources and manufacturing conditions during production, resulting in unique sulfur profiles for each brand of cheese. Since the precise lengths of maturation for each of these commercially obtained cheeses was unknown, some variation was expected to occur between labeled ages. Nevertheless, the quantitative results for the VSC content of these Cheddar cheeses, as seen in Figure 6, show that there is an increasing trend with age in the quantity of most VSCs.

The results from these commercial Cheddar cheeses show that there is a definite increase in the concentration of VSCs as the cheese matures. For example, those cheeses labeled as 'mild' contained the lowest overall concentrations of hydrogen sulfide, methanethiol, dimethyl sulfide, dimethyl disulfide and dimethyl trisulfide and those cheeses labeled as 'extra sharp' generally contained the highest concentrations. On average, hydrogen sulfide was present at the highest concentration, ranging from 43 parts per billion (ppb) in the mild cheese to 72 ppb in the extra sharp cheese. The average concentration range for dimethyl sulfide was 16 ppb in mild to 21 ppb in extra sharp; methanethiol: 7 ppb in mild to 14 ppb in extra sharp; dimethyl trisulfide: 1.5 ppb in mild to 2 ppb in extra sharp; and dimethyl disulfide: 0.4 ppb in mild to 0.6 ppb in extra sharp.

Volatile Sulfur Development in Cheeses made from Heat Shocked and Pasteurized Milk

Cheddar cheese that is made with pasteurized milk will often develop a less intense flavor, or take longer to do so, than Cheddar that is made with milk which has undergone less heat treatment. This is because the heat of pasteurization will kill off many native bacteria, including those bacteria that are beneficial for the flavor development. Six lots of Cheddar cheese were manufactured in a commercial plant; three lots were made from heat-shocked milk (66°C for 30 sec) and three lots from pasteurized milk (72°C for 16 sec) and all cheeses were aged simultaneously under the same conditions. The preliminary results from this cheese aging investigation (Figure 7) has shown that Cheddar cheese made with heat-shocked milk developed higher concentrations of methanethiol, dimethyl sulfide, dimethyl disulfide and dimethyl trisulfide. Only hydrogen sulfide appeared to be in higher concentrations in cheese made with pasteurized milk. This data shows that there is a definite difference in the VSC development for cheeses made with different heat treatments.

Conclusions

A sensitive headspace-SPME-GC-PFPD method has been developed for volatile sulfur analysis in cheese. Reactive methanethiol was successfully stabilized and no artifact formation was observed. Hydrogen sulfide, methanethiol, and dimethylsulfide were found to constitute the majority of the overall sulfur profile while dimethyl disulfide and dimethyl trisulfide were present in lesser amounts. Quantitative analysis showed that the concentration of certain sulfur compounds, namely hydrogen sulfide, methanethiol, and dimethyltrisulfide could be directly correlated with age. Cheeses that were aged for the longest amount of time contained the highest concentrations of VSCs. Controlled cheese aging study indicated that the VSCs developed differently in

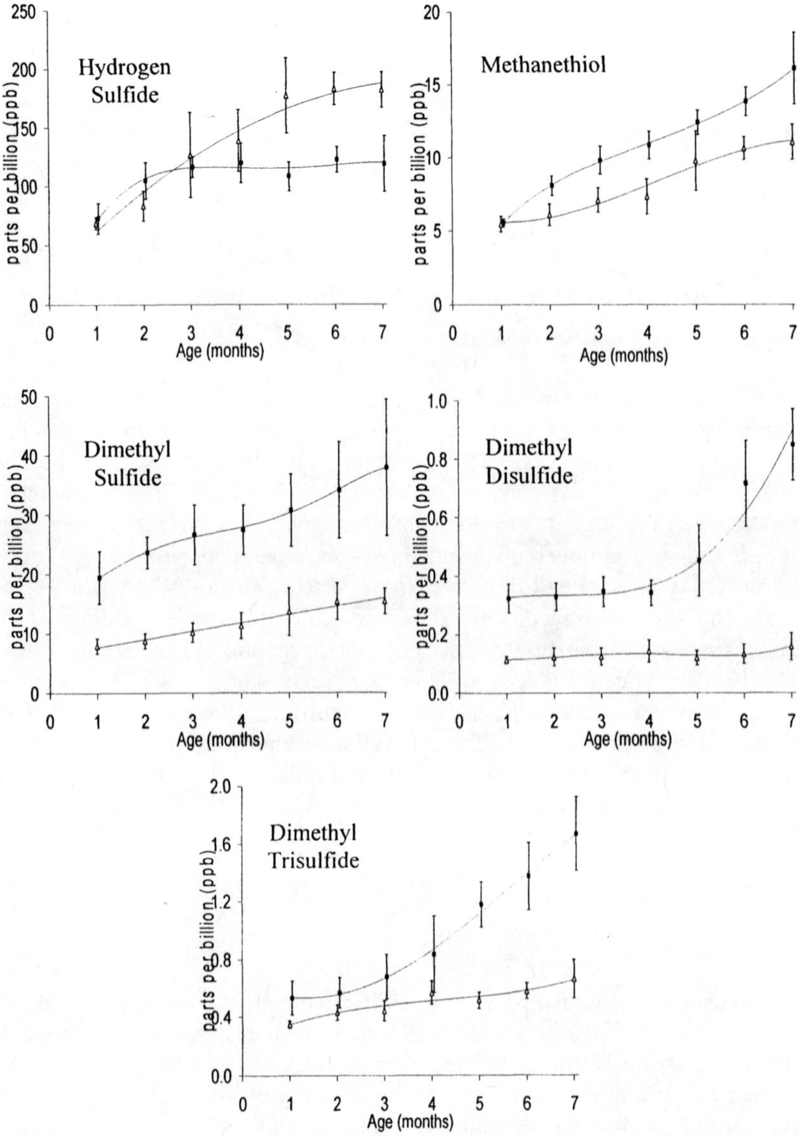

Figure 7. Trend of sulfur development during maturation of Cheddar cheese made with milk of two different heat treatments: heat-shocked (filled square, dashed line) versus pasteurized (open triangle, solid line). Error bars = deviation between three vats.

heat-shocked milk versus pasteurized milk during the aging process. Overall, the accurate quantification of VSCs in Cheddar provides some insight about the development of these important flavor compounds during maturation.

Literature Cited

1. Adda, J.; Gripon, J.C.; Vassal, L., *Food Chem.* **1982**, *9,* 115-129.
2. Preininger, M.; Rychlik, M.; Grosch, W., In *Trends in Flavor Research: Proceedings of the 7th Weurman Flavour Research Symposium*, Maarse, H; Van der Heij, D.G. (Eds), Elsevier Science: Amsterdam, **1994**, 267-270.
3. Kubícková, J.; Grosch, W., *Intl. Dairy J.* **1997**, *7*, 65-70.
4. Qian, M.; Nelson, C.; Bloomer, S., *J. Amer. Oil Chem. Soc.* **2002**, *79*, 663-667.
5. Qian, M.; Reineccius, G., *Flav. Fragr. J.* **2003**, *18,* 252-259.
6. Dias, B.; Weimer, B., *Appl. Environ. Microbiol.* **1998**, *64,* 3320-3326.
7. Dimos, A.; Urbach, G.E.; Miller, A.J., *Intl. Dairy J.* **1996**, *6,* 981-995.
8. Tanaka, H.; N, E.; Soda, K., *Enzyme Microb. Technol.* **1985**, *7*, 530-537.
9. Alting, A.C.; Engels, W.J.M.; Van Schalkwijk, S.; Exterkate, F.A., *Appl. Environ. Microbiol.* **1995**, *61,* 4037-4043.
10. Yuvon, M.; Thirouin, S.; Ruein, L.; Fromentier, D.; Gripon, J.C., *Appl. Environ. Microbiol.* **1997**, *63*, 414-419.
11. Gao, S.; Mooberry, E.S.; Steele, J.L., *Appl. Environ. Microbiol.* **1998**, *64*, 4670-4675.
12. Amarita, F.; Requena, T.; Taborda, G.; Amigo, L.; Pelaez, C., *J. Appl. Microbiol.* **2001**, *90*, 971-978.
13. Manning, D.J.; Moore, C., *J. Dairy Res.* **1979**, *46,* 539-546.
14. Manning, D.J., *J. Dairy Res.* **1974**, *41*, 81-87.
15. Manning, D.J.; Chapman, H.R.; Hosking, Z.D., *J. Dairy Res.* **1976**, *43*, 313-320.
16. Christensen, K.R.; Reineccius, G.A., *J. Food Sci.* **1995**, *60,* 218-220.
17. Milo, C.; Reineccius, G.A., *J. Agric. Food Chem.* **1997**, *45,* 3590-3594.
18. Singh, T.K.; Drake, M.A.; Cadwallader, K.R., *Comp. Rev. Food Sci. Food Safety* **2003**, *2,* 139-162.
19. Walker, J.R.L., *J. Dairy Res.* **1959**, *26,* 273-276.
20. Urbach, G., *Intl. Dairy J.* **1993**, *3,* 389-422.
21. Shankaranarayana, M.L.; Raghavan, B.; Abraham, K.O.; Natarajan, C.P., *CRC Crit. Rev. Food Tech.* **1974**, *4,* 395-435.
22. Grill, H.; Patton, S.; Cone, J.F., *J. Agric. Food Chem.* **1967**, *15,* 392-393.
23. Parliment, T.H.; Kolor, M.G.; Rizzo, D.J., *J. Agric. Food Chem.* **1982**, *30,* 1006-1008.

24. Burbank, H.M.; Qian, M.C., *J. Chromatogr. A* **2005**, *1066,* 149-157.
25. Fang, Y.; Qian, M.C., *J. Chromatogr. A* **2005**, *1080,* 177-185.
26. Chin, H.W.; Lindsay, R.C., *Food Chem.* **1994**, *49,* 387-392.
27. Preininger, M.; Grosch, W., *Lebensm.-Wiss. U.-Technol.* **1994**, *27,* 237-244.
28. Cuer, A., *Parfums, Cosmetiques, Aromes* **1982**, *44,* 88-92.
29. Meinhart, E.; Schreier, P., *Milchwissenschaft* **1986**, *41,* 689-691.
30. Wijesundera, C.; Urbach, G., *Final Report to the Dairy Research and Development Corporation, Project CSt 66 Melbourne* **1993**, 31.
31. Dumont, J.P.; Roger, S.; Adda, J., *Lait* **1976**, *56,* 595-599.
32. Muller, C.J.; Kepner, R.E.; Webb, A.D., *Am. J. Enol. Viticult.* **1971**, *22,* 156-160.
33. Dufour, J.P.; Delbecq, P.; Albela, L.P., In *Gas Chromatography-Olfactometry: the state of the art*, Leland, J.V.; Schieberle, P.; Buettner, A.; Acree, T.E. (Eds), American Chemical Society: Washington, D.C., **2001**, 123-137.
34. Libbey, L.M.; Bills, D.D.; Day, E.A., *J. Food Sci.* **1963**, *28,* 329-333.
35. Wilkes, J.G.; Conte, E.D.; Kim, Y.; Holcomb, M.; Sutherland, J.B.; Miller, D.W., *J. Chromatogr. A* **2000**, *880,* 3-33.
36. Nielsen, A.T.; Jonsson, S., *Analyst* **2002**, *127,* 1045-1049.
37. Hill, P.G.; Smith, R.M., *J. Chromatogr. A* **2000**, *872,* 203-213.
38. Lestremau, F.; Andersson, F.A.T.; Desauziers, V.; Fanlo, J.-L., *Anal. Chem.* **2003**, *75,* 2626-2632.
39. Roberts, D.D.; Pollien, P.; Milo, C., *J. Agric. Food Chem.* **2000**, *48,* 2430-2437.
40. Frank, D.C.; Owen, C.M.; Patterson, J., *Lebensm.-Wiss. U.-Technol.* **2004**, *37,* 139-154.
41. Lecanu, L.; Ducruet, V.; Jouquand, C.; Gratadoux, J.J.; Feigenbaum, A., *J. Agric. Food Chem.* **2002**, *50,* 3810-3817.
42. Pinho, O.; Peres, C.; Ferreira, I.M.P.L.V.O., *J. Chromatogr. A* **2003**, *1011,* 1-9.
43. Amirav, A.; Jing, H., *Anal. Chem.* **1995**, *67,* 3305-3318.
44. Titrimetric/potentiometric determination of H_2S, mercaptans, and COS in natural gas. *Metrohm Information* **2001**, *30,* 19-20.
45. Scheppers-Wercinski, S.A.; Pawliszyn, J., In *Solid-phase Microextraction: a practical guide,* Scheppers-Wercinski, S.A. (Ed), Marcel Dekker: New York, **1999**, 1-26.
46. Pinho, O.; Ferreira, I.M.P.L.V.O.; Ferreira, M.A., *Anal. Chem.* **2002**, *74,* 5199-5204.
47. Yang, X.; Peppard, T.L., In *Solid-phase Microextraction: a practical guide,* Scheppers- Wercinski, S.A. (Ed), Marcel Dekker: New York, **1999**, 177-201.

Chapter 9

Origins of Cheese Flavor

Michael H. Tunick

Dairy Processing and Products Research Unit, Eastern Regional Research Center, Agricultural Research Service, U.S. Department of Agriculture, 600 East Mermaid Lane, Wyndmoor, PA 19038

The flavor of cheese is affected by many factors, including milk quality stemming from the quality of the milk, processing parameters such as pasteurization and addition of salt, and enzymatic and chemical reactions that occur as the cheese ages. Lactose and citrate are metabolized by lactic acid bacteria to form a number of important compounds, including acetoin, 2,3-butanediol, and diacetyl, which generate buttery, cheesy flavors. Proteolysis of casein by coagulant, plasmin, and other enzymes leads to the production of acids, alcohols, aldehydes, amines, and amino acids, which bring about alcoholic, fatty, and green flavor notes. The breakdown of aromatic, branched-chain, and sulfur-containing amino acids also produces flavor compounds, many of which are undesirable. Triacylglycerols are lipolyzed into fatty acids, which impart pungent, cheesy flavors. Fatty acids can then be converted into methyl ketones, secondary alcohols, lactones, esters, and other compounds, which are responsible for earthy, floral, fruity, and rancid flavors. An array of compounds contributes to the unique flavor characteristics of each cheese variety.

The many types of cheese in the world provide an abundance of flavors. Starting with a bland product, milk, a cheesemaker adds starter cultures (selected lactic acid bacterial strains added to the milk), enzymes, and other chemicals (such as NaCl), and adjusts manufacturing procedures to produce any one of hundreds of cheese varieties, each with a characteristic flavor. Flavor in cheese arises from compounds in milk (transferred directly from diet, synthesized in mammary tissue, or added to the milk), control of manufacturing conditions (pasteurization, heat, etc.), and enzymatic and non-enzymatic chemical reactions. Flavors continue to form during storage, resulting from metabolism of lactate, lactose and citrate, proteolysis of caseins, and lipolysis of triacylglycerols. Compositional changes occur after milk is separated into curd and whey, the whey is drained off, and the curd is transformed into cheese. Cheddar cheese provides a representative example of these changes (Table I).

Table I. Percentages of components in bovine milk and fresh Cheddar cheese (*1, 2*).

	Lactose	*Citrate*	*Protein*	*Fat*
Milk	4.8	0.18	3.2	3.3
Cheese	0.8-1.0	0.2-0.5	25	33

Hundreds of compounds have been detected in cheese, many of which contribute to flavor (*3*), often in a synergistic manner. The purpose of this chapter is to review the processing parameters and the classes of compounds that lead to these flavors.

Processing Parameters

Processing parameters affecting cheese flavor include the condition of the milk, pasteurization, homogenization, starter cultures, non-starter lactic acid bacteria, coagulant, whey syneresis, salt, surface microorganisms, and ripening time, temperature, and humidity.

Milk

Changes in cows' diet will affect the flavor of milk and therefore the flavor of cheese. A common management practice that keeps feed consistent throughout the year is the use of silage, which consists of alfalfa, barley, corn, vetch, and wheat that has been allowed to ferment anaerobically and then fed to cows. In contrast, cows on pasture eat what they prefer. Some compounds from grasses, flowers, and other plants ingested by cows on pasture either pass directly into milk or are converted in the cow into other compounds (*4*). Many of these compounds impart desirable flowery flavors to cheese (*5*). When live pasture plants are consumed by cows, the plants activate the lipoxygenase system, which may be part of an injury response (*6*). Lipoxygenase activity breaks down lipids and carotenoids to products that affect cheese flavor (*5*). Nonanal, 2-nonenal, 2,4-decadienal, and methyl jasmonate are some of the compounds generated in this fashion from unsaturated fatty acids in plants. Citronellol, carvone, and geranyl acetate are precursors or degradation products of carotenoids (*5*). Linalool and α-pinene are terpenes often found in artisanal Alpine cheese (*7*). Research on the effects of pasture plants on cheese is in its infancy -- many of the world's cheese varieties are made from pasture-fed animals, but the flavor origins of these products and their concentrations in cheese have barely been examined. Urbach has reviewed the effects of feed on flavor (*4*). The health of the cows will also have an effect on cheese, as high somatic cell counts resulting from mastitis and poor nutrition will adversely affect coagulation and proteolysis (*8*).

Pasteurization and Homogenization

Pasteurizing milk above 75°C denatures whey proteins, primarily β-lactoglobulin, which liberates sulfhydryls and volatile sulfides, especially H_2S (*9*). These compounds give the milk a cooked flavor that can carry over to the cheese. Pasteurizing at 62°C for 30 min is enough to kill many indigenous microflora while not imparting a cooked flavor to the resulting cheese, but this length of time is longer and less economical than the more common 71°C for 15 s (*10*). Pasteurization at high temperatures also causes heat-induced interactions of whey proteins with casein, which changes the characteristics of proteolysis by limiting accessibility of proteases and peptidases to casein micelles (*11*). In addition, pasteurization inactivates indigenous microflora and several milk enzymes, causing the resulting cheese to contain fewer flavor compounds than raw milk cheese because of decreased proteolysis and lipolysis (*12*). Some enzymes survive pasteurization and are present in the milk at the beginning of cheesemaking. Nonstarter lactic acid bacteria may also enter the curd from the environment and increase the extent of proteolysis, often detrimentally (*14*). Plasmin, an enzyme found in milk and not inactivated by pasteurization, also contributes to proteolysis. Cheesemilk is not usually homogenized because the fat globules are reduced in size, which alters texture and functionality of the resulting cheese. Small, homogenized fat globules are subject to a higher level of lipolysis, which may lead to rancid flavors (*8*) as discussed below. Increased lipolysis may be desirable in some cases; Feta and Blue cheese are often homogenized, and cream is sometimes homogenized before adding to cheesemilk.

Starter Cultures

Starter cultures include lactococci, lactobacilli, and leuconostocs, and are specifically selected for their ability to ferment lactose to lactic acid. These cultures are added at the start of cheesemaking in order to lower the pH of the milk before the coagulation step. Adjunct cultures are added at the same time; these are additional starters intended to produce particular flavors and may include propionibacteria, which produces the openings in Swiss cheese, and *Penicillium roqueforti* mold, which is used in Blue and Brie manufacture. Lipase enzymes are added in the manufacture of Parmesan and Romano, enhancing lipolysis. Lactic acid bacteria influence the rate and extent of acid development, which controls the dissociation of colloidal calcium phosphate; these factors affect the rate of proteolysis and flavor production (*13*). Live starter cultures probably contribute little to flavor development once the cheese is made, but the enzymes they release upon cell death are responsible for proteolysis and some lipolysis (*8*).

Coagulant

Milk is coagulated by the addition of food-grade acid, which may impart its own flavor to cheese, or rennet, which is a mixture of proteolytic enzymes. Acid addition is not normally used for ripened cheese. Rennet, traditionally extracted from calf stomach, is now more economically and more commonly obtained from microbial sources such as *Rhizomucor miehei,* and from microbial fermentation. Chymosin, the primary enzyme in rennet, cleaves the κ-casein that stabilizes the casein micelle, causing the micelles to aggregate into a curd. Rennet survival in curd increases as the temperature of cooking decreases, leading to more proteolysis in the resulting cheese. Chymosin preferentially hydrolyzes α_{s1}-casein, but *R. miehei* enzymes also attack α_{s2}- and β-casein, which leads to greater peptide formation and additional flavors (*13*).

Syneresis

Whey imparts a tart flavor that may be objectionable in many cheese varieties. Around 90% of coagulated milk consists of whey, and cheesemakers try to remove much of it so they can isolate the curd. Syneresis (whey expulsion) decreases the moisture in the curd, which slows the growth of bacteria and therefore reduces proteolysis. Syneresis increases with the temperature and time at which the curd is coagulated and cooked. Cheesemakers also increase syneresis by reducing the size of the cubes into which the curd is cut, thus increasing their surface area (*15*). Whey proteins, which account for 20% of milk protein, comprise only 3-6% of the protein in cheese.

Salt and Coloring

Before cheese is packaged, NaCl is added to modify salty flavor and to regulate the growth of microorganisms. NaCl is added in dry form to the curd, or the cheese is immersed in brine. Starter culture activity and metabolism of lactose are inhibited when the salt-in-moisture concentration goes above 5% (*8*). Consequently, proteolysis, lipolysis, and flavor of cheese are affected by NaCl (*16*). Salt substitutes such as KCl are also used, but lead to uncharacteristic flavors. Addition of color such as annatto or β-carotene does not affect flavor, although some people believe that cheese tastes better when highly colored (*17*).

Surface Microorganisms

Cheeses such as Brick, Camembert, and Limburger are aged with the aid of mold, yeast, fungi, and coryneform bacteria on the surface. Molds include *Penicillium camemberti,* which metabolizes lactic acid (see Lactose section below) and triggers a series of reactions eventually resulting in the production of NH_3, leading to a pH of around 7.0 on the surface. The pH gradient (the interior of the cheese is initially around pH 4.6) causes migration of calcium phosphate to the surface and stimulates the action of the indigenous milk proteinase plasmin (*18*). Plasmin is responsible for some proteolysis and NH_3 imparts an ammonia flavor to cheese. Yeasts such as species of Kluyveromyces and Candida, and the *Geotrichum candidum* fungus, metabolize lactate to CO_2 and H_2O, providing an environment in which the coryneform bacteria *Brevibacterium linens* flourishes (*8*). *B. linens* and *G. candidum* break down sulfur-containing amino acids, leading to desirable sulfur-type flavors to surface-ripened cheese (*19*).

Ripening Time

As cheese undergoes aging (also known as curing, maturing, or ripening), various flavors become prominent and then may strengthen or fade because of the enzymatic processes taking place. These processes are accelerated by increasing the storage temperature, but most cheese is held at a typical temperature for that variety. For instance, Cheddar is ripened at 6-8°C and most Dutch and mold-ripened varieties are aged at 12-15°C. Ripening time, required for the desired flavor and texture, varies according to the cheese: Mozzarella needs only 3 wk, sharp Cheddar usually takes at least 6 mo, and Parmesan requires 2 yr (*3*). Soft cheeses are usually consumed within weeks because of bacterial activity. The storage times at which peak concentrations of common flavor compounds in Cheddar were reached during 26 wk of aging are compared in Table II. A combination of three different gas chromatography-olfactometry studies on Cheddar cheeses are represented in Table III -- a sharp cheese that was apparently under 8 mo old, a 1-yr-old cheese, and a 3-yr-old cheese. Acetic acid and 2,3-butanedione (diacetyl) were the only compounds found in all the cheeses in Table III, although methional was presumed to be present in the 1-yr-old cheese but lost in sample preparation (*21*). A compound responsible for mushroom flavor, 1-octen-3-one, was not positively identified in the 1-yr-old

Table II. Storage time at which flavor compounds in Cheddar cheeses reach their maximum concentration (*20*).

Compound	*Storage time (wk)*
Methanethiol	26
Hydrogen sulfide & dimethyl sulfide	26
Acetic acid	8-14
Butyric acid	20
Hexanoic acid	8-14
Octanoic acid	8-14
Decanoic acid	14
Ethanol	26
2-Propanol & 2-butanol	26
Acetone	14-20
Butanone	26
2,3-Butanedione	8
Heptan-2-one	20
Nonan-2-one	14-20
Undecan-2-one	14
δ-Decalactone	14
Ethyl acetate	4
Toluene	20

Table III. Flavor compounds in mild, 1-yr-old, and 3-yr-old Cheddar cheeses (*21, 23, 24*). An 'x' denotes a compound known to be present, and a '?' denotes one suspected of being present.

Compound	*Mild*	*1 yr*	*3 yr*
Acetaldehyde	x		
Methional	x	?	x
Propanal & 2-methylpropanal		x	
2-Methylbutanal		x	x
Pentanal			x
Hexanal	x	x	
Heptanal		x	
1- & 2-Nonanal	x		
trans-4,5-epoxy-2-decanal	x		
Methanethiol	x		
Carbon disulfide & dimethyl disulfide		x	
Dimethyl sulfide	x		
Dimethyl trisulfide & tetrasulfide	x		
Acetic acid	x	x	x
Butyric acid	x		x
Propionic, pentanoic, hexanoic acids			x
Decanoic & dodecanoic acids			x
Ethanol		x	
1-Proponol, 2-propanol, 2-methyl-1-propanol		x	
1-Butanol, 2-butanol, 1-pentanol		x	
2,3-Butanediol		x	x
2-Propanone, 2-butanone, 2-pentanone		x	
3-Hydroxy-2-butanone		x	x
2,3-Butanedione	x	x	x
Furanones[a]	x		
2-Heptanone		x	
1-Octen-3-one	x	?	x
β-Damascenone	x		
δ-Dodecalactone & 6-dodecen-γ-lactone	x		
Ethyl acetate, ethyl butyrate, ethyl hexanoate		x	x
Skatole	x		
α-Pinene			x
Toluene & various alkanes		x	

[a]4-Hydroxy-2,5-dimethyl-3(*2H*)-furanone and 2(5)-Ethyl-5(2)-methyl-4-hydroxy-3(*2H*)-furanone.

cheese but probably corresponded to a compound with a mushroom odor labeled as unknown. The laboratories involved used different procedures to separate and identify compounds, and the results may not be entirely comparable, but several trends are apparent. Some aldehydes and sulfur compounds are formed in young cheese and others are found later on, but most do not linger after 1 yr. The alcohols and most free fatty acids are not formed immediately. Lactones are formed early and then disappear, and esters are formed later on.

Few studies have monitored the formation of flavor compounds in one cheese during ripening. The results derived from one study that tracked volatile compounds in Serra da Estrela, an important Portuguese raw ewe's milk cheese, are shown in Table IV. The concentrations of the flavor compounds rise and fall because of the activity of the enzymes present (*22*).

Table IV. Changes in concentration (mg/kg cheese) of compounds in Serra da Estrela cheese during aging (*22*).

Compound	*Curd*	*3 wk*	*6 wk*
Butyric acid	20	47	37
Hexanoic acid	26	48	34
Octanoic acid	25	48	33
Decanoic acid	63	131	89
Dodecanoic acid	35	69	46
Total free fatty acids	626	1294	1012
Total alcohols (except methanol)	116	304	252
Total ketones (except diacetyl)	0.3	3.9	7.0

Lactose and Citrate

The breakdown products of proteins and lipids, the two principal organic constituents of cheese, are primarily responsible for cheese flavor. Although lactose and citrate are present at relatively small concentrations, their metabolism is also important in all cheese varieties (*8*).

Lactose

The principal carbohydrate found in milk is the disaccharide lactose; other carbohydrates occur only in trace amounts. Around 98% of the lactose in milk is lost with the whey during the manufacture of most cheeses. Residual lactose is completely metabolized by lactic acid bacteria to lactate within a day (up to 14 d for some dry-salted cheeses) (*8*), contributing an acidic flavor to cheese. Lactate serves as the primary source of food for starter bacteria that metabolize it to form acetaldehyde, acetate, ethanol, formate, and other compounds that contribute to characteristic cheese flavors (*25*). This process takes as little as 3 wk in Blue and other surface-ripened cheeses, and > 8 mo in Romano and other hard cheeses (*18*). CO_2 is also generated, forming eyes in the curd of Swiss and related varieties.

Citrate

Around 94% of the citrate in milk is in the soluble phase and is lost in the whey during cheesemaking (*26*). Although comprising < 1.0% of the curd (*1*), citrate is a major contributor to flavor (*26*). Citrate is co-metabolized with lactose to pyruvate, which is then broken down into 2,3-butanedione, 3-hydroxy-2-butanone (acetoin), 2,3-butanediol, and other compounds (*27*) (Fig. 1). These reactions occur during manufacture and ripening, and breakdown of the four-carbon metabolic products takes place as ripening continues. Citrate is still present at a concentration of 0.1% in Cheddar after 6 mo (*8*). 2,3-Butanedione is an important aroma and flavor compound for unripened, Cheddar, and Dutch-type cheeses (*17*) and is present throughout aging (Table III).

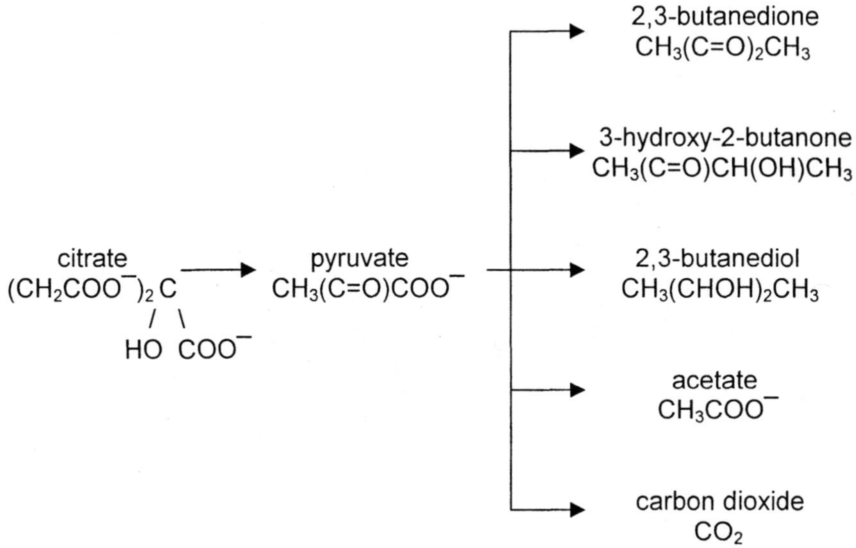

Figure 1. Breakdown of citrate in cheese.

Protein

Approximately 80% of cheese protein is casein, the proteolysis of which is the most complex and perhaps the most important biochemical event during cheese ripening (*1, 17*). Proteolysis of casein to peptides and amino acids results from action of the coagulant, plasmin, enzymes from starter and adjunct cultures, and enzymes from nonstarter bacteria. The process, which begins when coagulant is added and continues throughout ripening, is accelerated by increasing the moisture level in the cheese and the storage temperature, and by decreasing the salt content. Patterns of proteolysis are unique to each cheese variety; Mozzarella, with a ripening time of < 3 wk, exhibits low levels of proteolysis, but smear-ripened cheeses undergo extensive protein breakdown within weeks. The α_{s1}- and β-casein in blue-veined cheeses are completely hydrolyzed during ripening (*1*). Peptides are likely to contribute to the background flavor of cheese, but their role is uncertain. Larger peptides do not appear to have cheese-like flavors and small peptides are difficult to isolate and identify (*8*). Cheese proteolysis and analysis of proteins and peptides in cheese have been reviewed by Fox et al. (*25*) and by McSweeney and Sousa (*18*).

Amino Acids

Amino acids contribute to basic tastes in cheese: arginine, histidine, isoleucine, leucine, methionine, phenylalanine, tryptophan, tyrosine, and valine are bitter; alanine, glycine, serine, and threonine are sweet; lysine and proline are bitter and sweet; asparaginine, aspartic acid, and glutamic acid are sour; and sodium aspartate and sodium glutamate impart umami, or savory, flavor (*8*). Particular attention has been paid to bitter peptides, which are apparently released when coagulants and starter culture proteinases break the casein into fragments with hydrophobic ends (*25*). Proteolysis of β-casein by chymosin to the fragment β-casein (f193-209) has been implicated in bitterness (*3*). About half of the β-casein in Cheddar is hydrolyzed within 6 mo (*8*).

The most abundant amino acids in Cheddar are glutamic acid, leucine, lysine, phenylalanine, and valine (*28*), with methionine (see below) and leucine being the primary contributors to typical cheese flavor (*29*). Glutamic acid and leucine are present in Cheddar at approximately 100 mg/kg cheese after 5 wk, 300 mg/kg after 3 mo, and 650 mg/kg after 6 mo (*30*). Lysine, phenylalanine, and valine are present at 100-150 mg/kg after 3 mo and 200-350 mg/kg after 6 mo (*30*).

Amino acids act as precursors for catabolic reactions, as they undergo transamination to other amino acids, decarboxylation to amines, and deamination to α–ketoacids and ammonia (Fig. 2). In decarboxylation, an amino acid is converted to the corresponding amine with the loss of CO_2. Cheese contains numerous volatile amines such as dimethylamine, which is found in blue cheeses, and trimethylamine, which gives a fishy off-flavor. Primary and secondary amines usually have perception thresholds in the 1-200 mg/kg range, but tertiary amines such as trimethylamine can be perceived at 0.5 μg/kg (*31*). Volatile amines are responsible for alcoholic, fruity, and varnish flavors (*31*). Nonvolatile amines such as tyramine, histamine, and tryptamine are also generated, but these are more noted for their physiological effects on susceptible people than for flavor (*26*). Deamination leads to the production of α–ketoacids such as α–keto-3-methylbutanoic acid. Ammonia, which is important for the flavor of Camembert, Gruyère, and other cheeses, is also formed (*26*). Alkyl pyrazines, detected in some cheeses (*26*), contribute a roasted flavor (*32*).

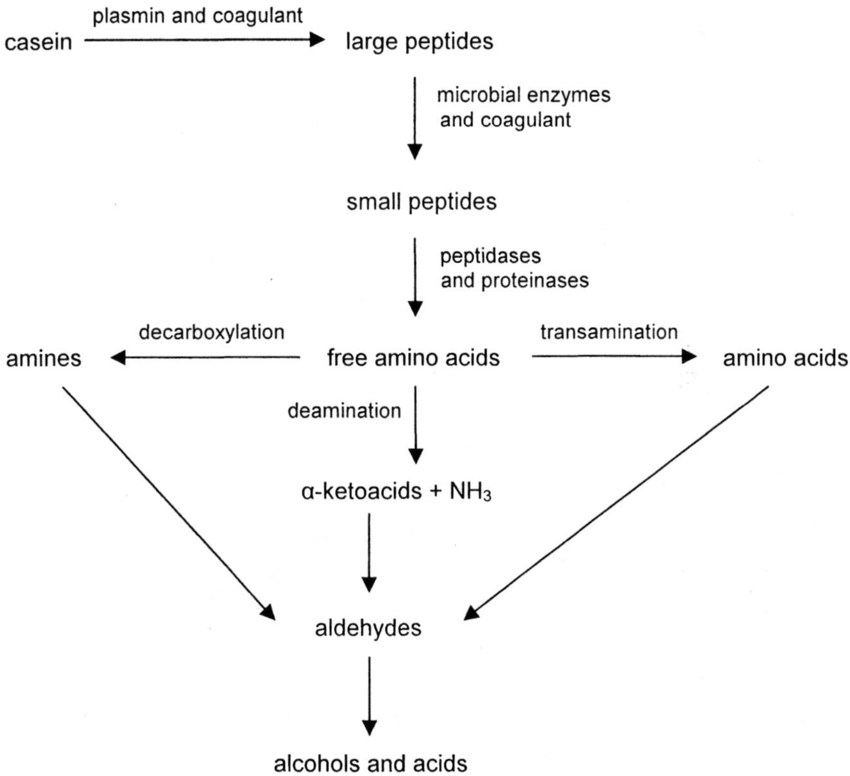

Figure 2. Proteolysis of casein into aldehydes, alcohols, and acids.

Aldehydes

Amines, amino acids, and α–ketoacids can be converted to aldehydes by the Strecker degradation, in which an α-amino acid reacts with a carbonyl group (*33*). Aldehydes are present throughout ripening (Table III). Perception thresholds of aldehydes are usually in the 1-10 μg/kg range (*31*). Saturated aldehydes from hexanal to nonanal lead to green immature fruit flavors (Table V), while longer chain aldehydes have aromatic notes (*31*). Isobutanal and 3-methylbutanal have recently been associated with nutty flavor (*34*). Aldehydes are often oxidized to acids or reduced to corresponding primary alcohols.

Table V. Selected aldehydes responsible for cheese flavors (*3, 32*).

Aldehyde	*Flavor notes*
Acetaldehyde	Sweet, pungent
2-Methylpropanal	Pungent, chocolate
Butanal	Pungent
2- & 3-Methylbutanal	Dark chocolate, malt
Pentanal	Pungent, almond-like
Hexanal	Green
2-Hexenal	Green, fatty, almond bitter
Heptanal	Green, fatty, oily
4-Heptenal	Creamy, biscuit
Octanal	Green, fatty, soapy, fruity
Nonanal	Green
2,6-Nonadieneal	Melon, cucumber
Decanal	Soapy, flowery
2,4-Decadienal	Mayonnaise, bread, fatty, fruity
Phenyl acetaldehyde	Rosy, honey

Methionine

Catabolism of sulfur-containing amino acids provides important flavor notes in many cheeses that are ripened for several months. Cysteine is present in trace amounts in cheese (*17*), but bacterial breakdown of methionine leads to potent compounds that are detectable in the 1-100 μg/kg range. The most important of these compounds is methanethiol (CH_3SH), which has been found to correlate with flavor intensity in Cheddar and is a precursor to sulfur compounds such as dimethylsulfide (*29*). Methional ($CH_3SCH_2CH_2CHO$), produced by the Strecker degradation, has a cooked potato odor that contributes in a positive manner to the flavors of Camembert, Cheddar, and other cheeses (*26*). Other sulfur-containing compounds resulting from methionine degradation include dimethyl-disulfide, dimethyltrisulfide, and various thioesters (*1, 19*). These compounds impart cabbage, cauliflower, garlic, and sulfurous flavor notes, and are especially important in mold-ripened cheeses (*31*).

Aromatic Amino Acids

Tryptophan, tyrosine, and phenylalanine are degraded into the α-ketoacids indole-3-pyruvate, *p*-hydroxyphenyl pyruvate, and phenyl pyruvate, respectively (*35*). From there, a number of pathways lead to important cheese flavor compounds, not all of which are desirable. Indole compounds and skatole produced from tryptophan have a fecal, putrid, musty odor (*36*). Benzaldehyde, produced from tryptophan and phenylalanine, has an almond aroma (*36*). Phenylethanol and phenylacetaldehyde, also produced from phenylalanine, have unclean flavors (*37*). Breakdown of tyrosine leads to the production of *p*-cresol, which is responsible for a barny or utensil flavor (*37*). All of these compounds are minimized by control of ripening temperature and time.

Branched Chain Amino Acids

Leucine, isoleucine, and valine are converted to α-ketoisocaproate, α-keto-β-methylvalerate, and α-ketoisovalerate, respectively, which lead to sweaty and rancid cheese flavors (*36*). The catabolic products 2-methylbutanal, 3-methylbutanal, and 2-methylpropanal impart malty flavors, and the corresponding alcohols lead to alcoholic and fruity odors (*38*).

Lipids

Oxidation of triacylglycerols in cheese is limited because of the negative redox potential, but hydrolysis by lipase enzymes to diacylglycerols, monoacylglycerols, glycerol, and fatty acids is extensive for many cheese varieties (*39*). Lipases in rennet paste coagulant, starter culture, and adjunct culture may all contribute to hydrolysis. Milk contains lipoprotein lipase, and additional lipase enzyme may also be added. The free fatty acids (FFA) level has been found to increase in Cheddar from 0.7 mmol/g cheese after 1 d, to 1.0 mmol/g after 60 d, to 2.5 mmol/g after 180 d; certain adjunct cultures can elevate these levels by 50% (*40*). The changes in FFA concentration in a raw milk cheese during aging are shown in Table IV.

Free Fatty Acids

Long-chain FFA with more than twelve carbon atoms, have high flavor thresholds (hundreds of mg/kg) and contribute slightly to cheese flavor. In contrast, FFA with shorter chains have far lower perception thresholds (1-10 mg/kg) and impart major flavor notes (*31*). FFA typically found in cheese are shown in Table VI. Some of these are directly responsible for flavors commonly associated with specific cheese varieties or cheese in general. In addition to the FFA listed, 2-methylbutyric, isobutyric, and isovaleric acids may be derived from branched-chain amino acids, and impart sweaty, rancid, and rotten fruit flavors (*36*). Hydrolytic rancidity, the release of flavorful short chain fatty acids by lipases, is undesirable if at high levels, but make a positive contribution to flavor at low levels, especially in hard Italian and Feta cheeses (*36*).

Fatty acids are catabolized into four classes of compounds: β-ketoacids, 4- and 5-hydroxyacids, unsaturated fatty acids, and free fatty acids (Fig. 3). Unsaturated fatty acids are degraded into aldehydes, acids, and alcohols, which were described in earlier sections.

β-Ketoacids

Methyl ketones and secondary alcohols are the degradation products of β-ketoacids. These appear at various times during ripening (Table III). Secondary alcohols (principally heptan-2-ol, nonan-2-ol, and phenyl-2-ethanol; Table VII) and odd-chain methyl ketones (especially 2-heptanone, 2-nonanone, and 2-undecanone; Table VIII) are responsible for a large portion of aroma compounds in mold-ripened cheese (*41*). The perception thresholds for these compounds range from 0.01 to 10.0 mg/kg (*31*). Citronellol and linalool originate in pasture plants (*5*). Two furan compounds, 2,5-dimethyl-4-hydroxy-3(2H)-furanone and 2-ethyl, 4-hydroxy-5-methyl-3(2H)-furanone, impart caramel flavor to Cheddar (*23*) and Swiss cheese (*42*).

Hydroxyacids

Hydroxyacids are esterified into cyclic compounds to form γ- and δ-lactones, which provide definite fruity notes (*31*). The most important of these compounds in cheese are the decalactones and dodecalactones, which can be perceived in the 0.1-3 mg/kg range (*31*) (Table IX).

Table VI. Selected fatty acids responsible for cheese flavors. (*3, 32, 41*).

Fatty acid	*Flavor notes*
Acetic & propionic	Pungent, vinegar
Butanoic	Rancid, cheese
n-Pentanoic	Swiss cheese
3-Methylbutanoic	Swiss cheese, waxy, sweaty, fecal
n-Hexanoic	Pungent, blue cheese, goat
n-Octanoic	Wax, soap, goat, musty, rancid, fruity, stale butter, sweaty
4-Methyloctanoic	Wax, goat, cheese
4-Ethyloctanoic	Goat
Decanoic	Rancid
Dodecanoic	Soapy

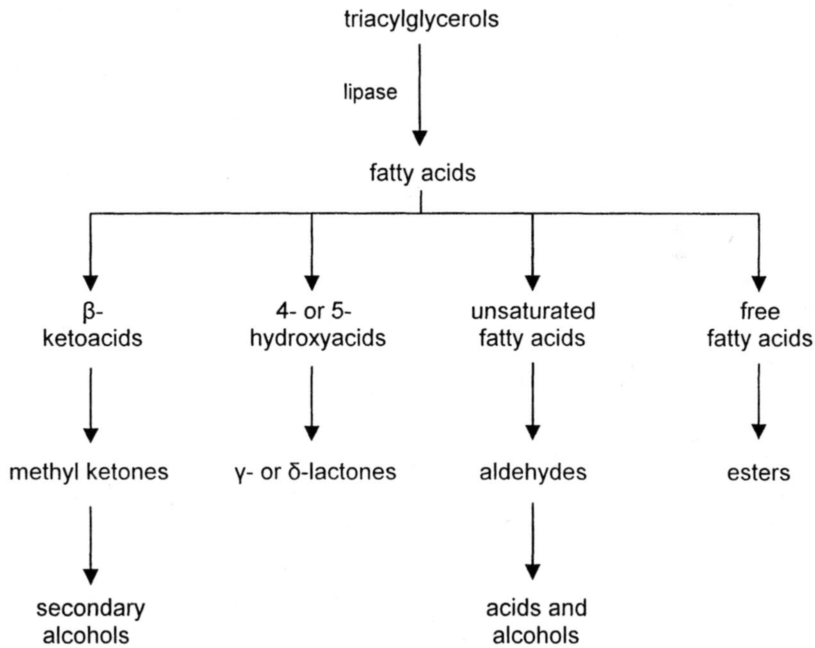

Figure 3. Lipolysis of triacylglycerols into cheese flavor compounds.

Table VII. Selected alcohols responsible for cheese flavors (*3, 32, 41*).

Alcohol	*Flavor notes*
Ethanol	Mild, ethery
n-Propanol	Pungent
n-Butanol	Floral, fragrant, fruity, sweet
2-Pentanol	Alcoholic, nutty, fruity, sweet
n-Hexanol	Fatty, floral, green
2-Octanol	Green, mushroom, coconut, oil, rancid
1-Octen-3-ol	Mushroom
1,5-Diocten-3-ol	Soil, geranium
Phenol	Floral, medical
2-Phenylethanol	Rose, floral
2-Methylisoborneol	Musty, soil
Citronellol	Rose
Linalool	Floral

Table VIII. Selected ketones responsible for cheese flavors (*3, 32, 41*).

Ketone	*Flavor notes*
2-Butanone	Etheric, acetone
2,3-Butanedione	Buttery
3-Methyl-2-butanone	Camphor
2-Heptanone	Blue cheese, fruity, musty, soapy
1-Octen-3-one	Mushroom
2-Octanone	Floral, fruity, musty, soapy
2-Nonanone	Green, earthy, blue cheese, fatty, fruity, musty, varnish
2-Undecanone	Floral, fruity, green, musty, tallow
Acetophenone	Orange blossom, almond, musty, glue
Carvone	Herbaceous

Table IX. Fruit-type flavors arising from lactones found in cheese (*31, 41*).

Lactone	*Flavor notes*
δ-Octalactone	Coconut, wine
γ-Octalactone	Fruity, coconut
δ-Decalactone	Peach, coconut
γ-Decalactone	Peach, apricot, coconut
δ-Dodecalactone	Fresh fruit, peach, pear, plum, coconut
γ-Dodecalactone	Peach

Esters

A variety of enzymes causes short- and medium-chain fatty acids to be esterified with ethanol to form esters, usually later in the ripening process (Table III). Esters have characteristic fruity or floral notes (Table X) which are often ten times more potent than the corresponding alcohols (*41*). Esters can be perceived in a range from 2 μg/kg to 20 mg/kg (*31*). Thioesters derived from methionine are also found in some cheeses, and are responsible for cabbage, cheesy, and rancid flavors (*43*).

Table X. Fruity and floral flavor notes arising from esters found in cheese (*3, 41*).

Ester	*Flavor notes*
Ethyl acetate	Fruity, sweet, solvent, pineapple
n-Butyl acetate	Pear
3-Methylbutyl acetate	Banana
Isoamylacetate	Pear, banana
Ethyl propionate	Fruity, pineapple
Ethyl butanoate	Fruity, pineapple, bubble gum
n-Propyl butanoate	Pineapple, banana
Methyl hexanoate	Pineapple
Ethyl hexanoate	Fruity, pineapple, banana
Ethyl octanoate	Fruity, apricot, wine
Geranyl acetate	Rose
Methyl jasmonate	Jasmine

Summary

Cheese acquires its unique mixture of flavors from variations of manufacturing parameters and from the breakdown of citrate, lactose, lactate, lipids, and proteins. Rather than being a stable and invariable product, cheese is constantly undergoing changes that affect its flavor profile.

References

1. McSweeney, P.L.H. *Int. J. Dairy Technol.* **2004**, *57*, 127-144.
2. U.S. Department of Agriculture, Agricultural Research Service. *USDA National Nutrient Database for Standard Reference,* Release 15, 2002. Nutrient Data Laboratory Home Page, www.nal.usda.gov/fnic/foodcomp
3. Singh, T.K.; Drake, M.A.; Cadwallader, K.R. *Compr. Revs. Food Sci. Safety* **2003**, *2*, 139-162.
4. Urbach, G. *J. Dairy Sci.* **1990**, *73*, 3639-3650.
5. Carpino, S.; Mallia, S.; La Terra, S.; Melilli, C.; Licitra, G.; Acree, T.E.; Barbano, D.M.; Van Soest, P.J. *J. Dairy Sci.* **2004**, *87*, 816-830.
6. Galliard, T.; Chan, H.W.-S. In *The Biochemistry of Plants. Lipids: Structure and Function;* Stumpf, P.K., Ed.; Academic Press: New York, 1980; Vol. 4, pp 131-161.
7. Curioni, P.M.G.; Bosset, J.O. *Int. Dairy J.* **2002**, *12*, 959-984.
8. Fox, P.F.; Guinee, T.P.; Cogan, T.M.; McSweeney, P.L.H. In *Fundamentals of Cheese Science;* Aspen Publ.: Gaithersburg, MD, 2000.
9. Calvo, M.M.; de la Hoz, L. *Int. Dairy J.* **1992**, *2*, 69-81.
10. Jenkins, S. *Cheese Primer;* Workman Publ.: New York, 1996, pp 2-3.
11. Lau, K.Y.; Barbano, D.M.; Rasmussen, R.R. *J. Dairy Sci.* **1991**, *74*, 727-740.
12. Grappin, R.; Beuvier, E. *Int. Dairy J.* **1997**, *7*, 751-761.
13. Nauth, K.R.; Hynes, J.T.; Harris, R.D. In *Encyclopedia of Food Science and Technology,* Hui, Y.H., Ed.; John Wiley & Sons: New York, 1992; Vol. 1, pp 333-348.
14. Morales, P.; Fernández-García, E.; Nuñez, M. *J. Agric. Food Chem.* **2005**, *53*, 6835-6843.
15. Hill, A.R. In *Chemistry of Structure-Function Relationships in Cheese;* Malin, E.L.; Tunick, M.H., Eds.; Plenum Press: New York, 1995; pp 43-58.
16. Guinee, T.P. *Dairy Ind. Int.* **2004**, *69*(12), 36-38.
17. Dunn, M.L.; Ogden, L.V. *J. Dairy Sci.* **1991**, *74*, 2042-2047.
18. Fox, P.F.; Singh, T.K.; McSweeney, P.L.H. In *Chemistry of Structure-Function Relationships in Cheese;* Malin, E.L.; Tunick, M.H., Eds.; Plenum Press: New York, 1995; pp 59-98.

19. Berger, C.; Khan, J.A.; Molimard, P.; Martin, N.; Spinnler, H.E. *Appl. Environ. Microbiol.* **1999**, *65*, 5510-5514.
20. Dimos, A.; Urbach, G.E.; Miller, A.J. *Int. Dairy J.* **1996**, *6*, 981-995.
21. Arora, G.; Cormier, F.; Lee, B. *J. Agric. Food Chem.* **1995**, *43*, 748-752.
22. Partidário, A.M.; Barbosa, M.; Vilas Boas, L. *Int. Dairy J.* **1998**, *8*, 873-881.
23. Milo, C.; Reineccius, G. *J. Agric. Food Chem.* **1997**, *45*, 3590-3594.
24. Christensen, K.R.; Reineccius, G.A. *J. Food Sci.* **1995**, *60*, 218-220.
25. McSweeney, P.L.H.; Sousa, M.J. *Lait* **2000**, *80*, 293-324.
26. Hugenholtz, J. *FEMS Microbiol. Revs.* **1993**, *12*, 165-178.
27. LeQuéré, J.-L.; Molimard, P. In *Encyclopedia of Dairy Sciences*; Roginski, H.; Fuquay, J.W.; Fox, P.F., Eds.; Academic Press: London, 2002; Vol. 1, pp 330-340.
28. Law, B.A.; Castañón, M.J.; Sharpe, M.E. *J. Dairy Res.* **1978**, *45*, 267-275.
29. Aston, J.W.; Creamer, L.K. *New Zeal. J. Dairy Sci.* **1986**, *21*, 229-248.
30. Wallace, J.M.; Fox, P.F. *Int. Dairy J.* **1997**, *7*, 157-167.
31. Acree, T.; Arn, H. *Flavornet*, 2004. Flavornet Home Page, www.flavornet.org/index.html
32. Fox, P.F.; McSweeney, P.L.H.; Cogan, T.M.; Guinee, T.P. In *Cheese: Chemistry, Physics and Microbiology;* Academic Press: London, 2004; 3rd edn., Vol. 1; pp 361-371.
33. Weenen, H.; van der Ven, J.G.M. In *Aroma Active Compounds in Foods;* Takeoka, G.R.; Güntert, M.; Engel, K.-H., Eds.; ACS: Washington, DC, 2001; pp 183-195.
34. Avsar, Y.K.; Karagul-Yuceer, Y.; Drake, M.A.; Singh, T.K.; Yoon, Y.; Cadwallader, K.R. *J. Dairy Sci.* **2004**, *87*, 1999-2010.
35. Curtin, A.C.; McSweeney, P.L.H. In *Cheese: Chemistry, Physics and Microbiology;* Fox, P.F.; McSweeney, P.L.H.; Cogan, T.M.; Guinee, T.P., Eds.; Academic Press: London, 2004; 3rd edn., Vol. 1; pp 435-454.
36. Marilley, L.; Casey, M.G. *Int. J. Food Microbiol.* **2004**, *90*, 139-159.
37. Dunn, H.C.; Lindsay, R.C. *J. Dairy Sci.* **1985**, *68*, 2859-2874.
38. Thierry, A.; Maillard, M.-B. *Lait* **2002**, *82*, 17-32.
39. Collins, Y.F.; McSweeney, P.L.H.; Wilkinson, M.G. *Int. Dairy J.* **2003**, *13*, 841-866.
40. Madkor, S.A.; Tong, P.S.; El Soda, M. *J. Dairy Sci.* **2000**, *83*, 1684-1691.
41. Molimard, P.; Spinnler, H.P. *J. Dairy Sci.* **1996**, *79*, 169-184.
42. Preininger, M.; Grosch, W. *Lebensm. Wiss. Technol.* **1994**, *27*, 237-244.
43. Berger, C.; Martin, N.; Collin, S.; Gijs, L.; Khan, J.A.; Piraprez, G.; Spinnler, H.E.; Vulfson, E.N. *J. Agric. Food Chem.* **1999**, *47*, 3274-3279.

Biochemical Aspects

Chapter 10

Biochemistry of Cheese Flavor Development: Insights from Genomic Studies of Lactic Acid Bacteria

Jeffery R. Broadbent[1] and James L. Steele[2]

[1]Department of Nutrition and Food Sciences and Western Dairy Center, Utah State University, Logan, UT 84322
[2]Department of Food Science, University of Wisconsin, Madison, WI 53706

To attain characteristic flavor and body attributes, many cheese varieties must be ripened at low temperature for months or even years. Because flavorful cheese has premium value as a food or food ingredient, there is considerable industrial interest in technologies to accelerate the ripening process. Research has shown that the lactic acid bacteria (LAB) present in the cheese matrix have a central role in flavor development, so effective strategies to accelerate or intensify cheese flavor can be derived from a more fundamental understanding of LAB physiology in milk and cheese environments. Although many details must still be elucidated, current knowledge indicates LAB influence cheese flavor development via several key mechanisms including proteolysis, amino acid metabolism, lipase/esterase activity and citrate utilization. This paper provides an overview of these reactions and addresses recent genomic-based advancements that are driving current research to understand how LAB influence cheese flavor development.

Long before the emergence of microbiological science, human civilizations developed effective technologies to enhance the keeping qualities of foods, and one of the oldest methods for food preservation involves fermentation by lactic acid bacteria (LAB). The LAB are a relatively heterogeneous group of gram-positive cocci, coccobacilli, and bacilli that inhabit a variety of ecological niches, yet share several defining characteristics such as: 1) low (< 55 mol%) G + C content; 2) high acid tolerance; 3) nonsporing; 4) nutritionally fastidious; 5) aerotolerant but not aerobic; 6) inability to synthesize porphyrins; and 7) a strictly fermentative metabolism with lactic acid as the major metabolic end product (*1*). This group includes several industrially important species of *Lactobacillus* (*Lb.*), *Lactococcus* (*Lc.*), *Leuconostoc* (*Ln.*), and *Streptococcus*, which serve as starter cultures for the commercial manufacture of cheese and fermented milks. Because these bacteria are natural constituents of raw milk, spontaneously produced fermented milks have likely been part of the human diet for millennia. In time, such inadvertent fermentations slowly evolved into more deliberate processes that today provide more than 1000 unique cheeses, yogurts, and fermented milks. Historically, manufacturing processes for cheese and other fermented dairy foods relied entirely upon spontaneous acidification of milk by naturally occurring LAB. However, the discovery of the lactic acid fermentation by Pasteur in the mid-19th century, and the isolation of LAB starter cultures that quickly followed, eventually propelled many of these artisan processes into large-scale industrialized fermentations.

In its continual drive toward increased mechanization and throughput, the dairy industry has become a leader in microbial and fermentation technologies. This is because experience has demonstrated that industrial-scale production of flavorful, consistently high-quality cheese is dependent on the use of well-characterized starter bacteria. Because growth and economic success in cheese industry still depends on starter cultures with defined, predictable, and stable properties, fundamental knowledge of LAB genetics and physiology holds great value to the world's milk fermentation industry (*2*).

Cheese Microbiology and Flavor Development

Transformation of bland-flavored curd into delicious mature cheese is a complex and dynamic process whose intricacies are scripted by the milk type and composition, the cultures and enzymes present or added to the cheese milk, and the manufacturing and ripening conditions used. Many cheeses need to be stored at low temperature for months or even years before they attain characteristic flavor and body attributes. During this time, termed the curing or ripening period, microorganisms and enzymes in the cheese matrix act on milk constituents in a manner that is partly dictated by the curd microenvironment (e.g., cheese pH, a_w, salt content, E_h, temperature, etc.) and which ultimately gives the desired product (*3*).

Types of LAB that occur in internally bacterial-ripened cheeses such as Cheddar, Dutch, Swiss, and Italian varieties, include deliberately added strains (e.g., starters and adjunct cultures) and adventitious species (primarily nonstarter LAB or NSLAB) that enter cheese through milk or processing equipment. Modern sanitation and Good Manufacturing Practices help minimize initial numbers of NSLAB in cheese, yet these organisms invariably appear and grow to high numbers during ripening (*4*). In Cheddar cheese, numbers of *Lactococcus lactis* starter bacteria frequently exceed 10^9 colony-forming units (cfu) per gram when ripening begins. As maturation proceeds, the harsh cheese ripening environment (little or no residual lactose, pH 5.0 to 5.3, 4-6% salt in moisture, 5-13°C) gradually takes its toll and starter viability declines. A fraction of the dying starter cells undergo autolysis, which releases intracellular enzymes and cellular components (e.g., sugars and nucleic acids) into the cheese matrix (*5*). At the same time, NSLAB populations (whose initial numbers are typically less than 10^2 cfu/g in cheese made under good sanitary conditions with high quality pasteurized milk) begin to grow and eventually plateau at cell densities of 10^7-10^8 cfu/gram after 3-9 mo of aging (*4*). Microbiological studies have shown NSLAB populations in bacterial-ripened cheeses may be quite diverse, but are usually dominated by facultatively heterofermentative species of lactobacilli or, far less frequently, by pediococci (*4,6-10*). Depending on the species that is used (and whether or not the particular strain can grow in ripening cheese), populations of adjunct bacteria may mirror the trend for starter or NSLAB fractions.

Though a link between LAB activity and cheese flavor attributes was postulated more than 100 years ago (see ref *5*), the variation and complexity that exists in cheese microbiota and enzyme content confounded early efforts to establish a causal role for these bacteria in flavor development. This limitation was overcome in the late-1950's, when sensory studies of aseptically manufactured Cheddar cheese showed that starter-free, gluconolactone-acidified cheese failed to develop Cheddar flavor, while cheese made with *Lc. lactis* starter bacteria developed characteristic, balanced flavor (*11,12*). The same investigations also showed NSLAB could modify basic flavor notes and accelerate flavor development. More recently, use of *Lactobacillus* spp. NSLAB isolates as adjunct cultures for Cheddar cheese manufacture has indicated these bacteria may influence flavor in at least 3 ways: they may intensify (i.e., accelerate) typical flavor development, impart atypical (but desirable) flavor notes, or promote off-flavor development (*5, 9, 10, 13, 14, 15, 16*). In addition, NSLAB have also been associated with cheese quality defects such as open body (via gas production) and formation of calcium lactate crystals (*5, 17, 18*).

LAB Physiology and Cheese Flavor Development

Given the causal role of LAB in flavor development, efforts to define the biochemical basis for flavor changes in cheese have logically focused on the microbiology and physiology of species found in cheese (for recent reviews see refs *6,19-22*). Those efforts have found starter, adjunct, and NSLAB can influence flavor development through several mechanisms that include lactose fermentation, conversion of milk proteins (primarily caseins) into peptides and free amino acids, catabolism of amino acids into volatile aroma compounds, lipase/esterase activity, and citrate catabolism.

Fermentation of lactose to L-lactic acid is a primary function of any starter culture in cheese manufacture, and the degree of acid production in the vat is a critical factor for controlling cheese quality. This is because the rate of acid production by the culture determines the final pH and mineral content of the curd, which in turn affects the basic protein structure and amount of residual coagulant retained in the curd, and thus cheese texture and flavor properties (23). Lactate itself is a component of cheese flavor and in Swiss-type cheeses, it serves as a substrate for the growth of propionibacteria, which convert lactate into propionic acid (another important flavor component) and CO_2 (which produces the cheese "eyes"). Rapid depletion of residual milk sugar in the curd by starter bacteria also helps prevent its use as a substrate for undesirable adventitious bacteria that may produce flavor and body defects (*5,24*).

In many bacterial-ripened cheeses, the enzymatic conversion of caseins into peptides and free amino acids is one of the most important biochemical events during ripening (*3*). As a result, the relationship between proteolysis and cheese flavor development has been an area of intense research interest for decades. That work has revealed that proteolysis in ripened cheese is a complex process that involves endogenous milk enzymes, chymosin, and microbial proteinases and peptidases (*25*). The relative contribution of these enzymes to cheese ripening is influenced by the specificity and relative activity of individual enzymes in the cheese matrix and, in the case of intracellular enzymes, access to appropriate substrate (*26*). In ripening Cheddar cheese, the hydrolysis of intact caseins is almost exclusively catalyzed by chymosin and endogenous milk enzymes, while the proteinases and peptidases from *Lc. lactis* starter bacteria are primarily responsible for the production of water-soluble peptides and free amino acids (*27*).

Together, primary and secondary proteolysis may influence cheese flavor in several ways (*3*). For example, breakdown of the casein network by chymosin and plasmin loosens cheese texture, which facilitates the release of flavor compounds when the cheese is consumed. In addition, some of the peptides produced by these enzymes can have a direct influence on flavor, but that effect is probably negative (i.e., bitter flavor defect). The high- and medium-molecular weight peptides produced by chymosin and plasmin also serve as substrates for microbial proteinases and peptidases, which convert them into low molecular-weight peptides and free amino acids. Free amino acids can also have direct effects on cheese flavor (e.g., sweet, sour, bitter, or umami flavors; *20*), or may serve as precursors for the production of a broad range of potent aroma compounds (*28*).

The latter reactions are of particular interest because a growing body of evidence indicates that amino acid catabolism by LAB is a key mechanism by which these bacteria generate flavor compounds in cheese (*29*). Thus, pathways for amino acid catabolism in LAB and their impact on cheese flavor development has been the subject of several recent reviews (*21,26,28,30,31*). The products of amino acid catabolism, which may arise via decarboxylation, deamination, transamination, desulfuration, or side chain removal, can impart positive or negative attributes to cheese flavor. To date, much of the research on amino acid catabolism by LAB has been directed toward the fates of aromatic, sulfur-containing, and branched chain classes of amino acids because of their key role in aroma development (*28*). As examples, Met conversion contributes desirable "sulfur" flavor to many cheese types (*31*), Leu catabolism is a source for "nutty" flavor in Cheddar cheese (*32*), and Phe breakdown can generate "floral" notes or "unclean" flavors (*33,34*).

Lipase or esterase activity can also have direct and indirect consequences on cheese flavor. Enzymes involved in these reactions may come from milk itself, be added, or arise from starter and nonstarter LAB. Lipase addition, for example, produces the short-chain free fatty acids responsible for sharp flavor notes in some Italian cheeses (*35*). In cheeses made from pasteurized milk without added lipase or esterase, flavors associated with lipolytic or esterolytic activity are likely due to milk or microbial enzymes (*35*). Fatty acids released by lipase or esterase activity on milk triglycerides also serve as precursors for the formation of esters that impart fruity flavor to cheese. Interestingly, esterases and lipases are able to catalyze the hydrolysis and synthesis of esters, depending on conditions and cosubstrate levels (i.e., fatty acids and alcohols) (*36,37*). Most LAB have very low lipolytic or esterolytic activity, but these organisms can generate sufficient levels of free fatty acids and esters to impact flavor in cheese with long ripening times (*37,38*).

Finally, citrate utilization by LAB is important to flavor because it can lead to the production of succinate, a compound with monosodium glutamate-like flavor enhancing properties, or the butter flavor compound, diacetyl. In Cheddar cheese, succinate is most likely produced by NSLAB from citrate via the reductive TCA pathway (*39*). The importance of diacetyl to the flavor and aroma of butter, buttermilk, and many cheeses has been recognized for decades. Diacetyl is formed by oxidative decomposition of α-acetolactate, an intermediate in the pathways for pyruvate metabolism and amino acid biosynthesis. In recent years, detailed knowledge of citrate metabolism and diacetyl production has yielded effective strategies to metabolically engineer *Lc. lactis* strains for enhanced diacetyl production (*40*).

Greater knowledge of the mechanisms by which LAB affect cheese flavor has facilitated industry efforts to accelerate or intensify cheese flavor development but in many cases, detailed understanding of the molecular processes that direct flavor production is still lacking. At present, most of the key advances in this area are derived from experiments based on recombinant DNA technology. The advantage this technology affords for the analysis of LAB cellular and industrial processes is that it facilitates construction of strain derivatives that differ only by the action (null mutants) or activity (expression mutants) of one or more defined polypeptides. By comparing the phenotype of the wild-type bacterium to its isogenic derivative, the role of that polypeptide in a given process can often be defined. The knowledge accumulated from this type of research ultimately serves efforts to isolate or construct new strains with enhanced industrial productivity. As an example, the complexity of the LAB peptidase enzyme system had, for many years, confounded efforts to determine the role of individual enzymes in cheese proteolysis and maturation. However, construction of isogenic peptidase mutants now provides researchers with a potent systematic approach to determine the contribution of individual enzymes on cell growth and cheese properties (*26*,*41*,*42*,*43*).

From Genetics to Genomics

Because of its industrial significance as the starter bacterium for manufacture of Cheddar and Dutch-type cheeses, and the ease by which it may be propagated in the laboratory, current understanding of genetics in "food-grade" LAB has largely been driven by research on *Lc. lactis*. Genetic studies of *Lc. lactis* began in the early 1970's and since that time, four basic groups of genetic elements have been characterized to the nucleotide sequence level in this and other dairy LAB species: plasmid DNA, mobile DNA elements, bacteriophages and, most recently, entire chromosomes (*2*). Though representatives from all four groups have been shown to affect milk fermentation, recent advances in fundamental knowledge of chromosomal structure and organization are of particular significance because they outline all of the essential housekeeping, catabolic, and anabolic activities of the cell.

As outlined in a recent review by Broadbent (*2*), efforts to characterize LAB chromosomes were begun in the early 1970s, but detailed studies were not feasible until the development of pulsed-electric field gel electrophoresis (PFGE) technology in the early 1980's. With this method, researchers were able to determine LAB have a relatively small (1.8 to 3.4 Mbp), single, and circular chromosome (e.g., fig. 1), and that genome size and organization differed among individual species and even some strains (*44,45*). The PFGE method is still useful in chromosome studies, but the most exciting developments in LAB genomics are now being fueled by nucleotide sequence information for complete genomes.

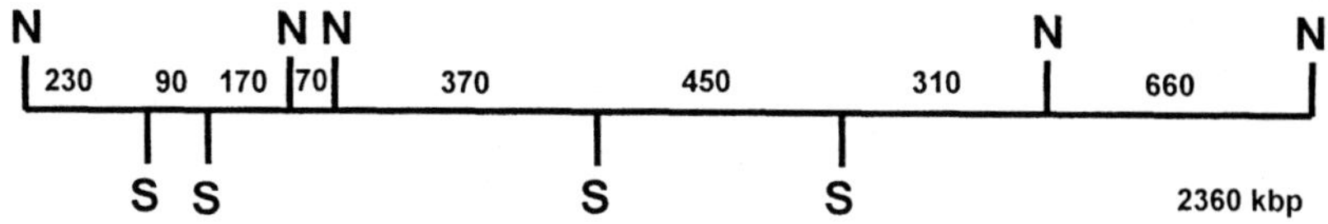

Figure 1. Physical map of the Lactobillus helveticus CNRZ32 chromosome. The map was derived from data collected after pulse-field gel electrophoresis with the restriction enzymes NotI (N) and SfiI (S). Numbers represent fragment sizes in kilobase pairs (kbp)

Compilation and annotation (software-assisted prediction of gene location and product function) of full genome sequences has revolutionized bacteriology and provided exciting new opportunities to investigate bacterial evolution, genetics, physiology, and metabolism. As is shown in Table I, genome sequence information for several important LAB has become available in recent years, and sequencing efforts are underway for many other species that have significance in milk fermentation (*46*). In fact, the number of sequencing projects for LAB has become extensive enough that sequence information is now available for more than one strain in some industrially important species (Table I). While the latter development may at first glance appear to embody a redundant expenditure of sequencing resources, it is actually an important and exciting outcome. This is because it should allow researchers to elucidate the molecular basis for a number of important strain-dependent properties that are commonly encountered in dairy LAB, such as the ability to produce specific flavor notes, rates for autolysis or acidification, and resistance to freezing or lyophilization.

Though some of the genomes listed in Table I are not yet available to the general scientific community, sequence data is now accessible for more than half of the strains listed. It is noteworthy that ten of the publicly accessible sequences listed in Table I were derived from a joint venture between the Department of Energy's Joint Genome Institute and the Lactic Acid Bacterial Genomics Consortium (LABGC). The goal of this Consortium is to advance academic and industrial research on LAB through compilation of publicly accessible genome sequence data for microbes that are important to the fermented foods industry. For more information on the LABGC effort see:

Table I. Completed and Ongoing Genome Sequence Projects for Bacteria Relevant to Milk Fermentation

Species	*Genome size (Mbp)*	*Project sponsor*[1]	*Data access?*[1]
Lactic acid bacteria:			
Lactobacillus acidophilus ATCC 700396	2.0	Dairy Management, Inc./Rhodia, Inc. (USA)	yes
Lb. brevis ATCC 367	2.0	JGI-LABGC (USA)	yes
Lb. casei ATCC 334	2.9	JGI-LABGC (USA)	yes
Lb. casei BL23	2.6	INRA (FRA)	no
Lb. delbrueckii subsp. *bulgarius* ATCC BAA-365	2.3	JGI-LABGC (USA)	yes
Lb. delbrueckii subsp. *bulgarius* ATCC 11842	2.3	INRA / Genoscope (FRA)	no
Lb. delbrueckii subsp. *bulgarius* DN-100107	2.1	Danone Vitapole (FRA)	no
Lb. gasseri ATCC 33323	2.0	JGI-LABGC (USA)	yes
Lb. helveticus CM4	2.0	Calpis Co. / Kitasato Univ. (JPN)	no
Lb. helveticus CNRZ 32	2.4	Dairy Mngmnt Inc./Chr. Hansen, Inc. (USA)	no
Lb. helveticus DPC 4571	2.1	Teagasc / Univeristy College, Cork (IRL)	no
Lb. johnsonii NCC533	2.0	Nestlé (CHE)	yes
Lb. plantarum WCFS1	3.3	Wageningen Centre for Food Science (NLD)	yes
Lb. reuteri 100-23	NR	Univ. Otago (NZ) / JGI (USA)	no
Lb. reuteri DSM20016	2.4	Univ. Otago (NZ) / JGI (USA)	no
Lb. rhamnosus HN001	2.3	Fontera Research Center (NZ)	no
Lactococcus lactis subsp. *cremoris* SK11	2.6	JGI-LABGC (USA)	yes
Lc. lactis subsp. *cremoris* MG1363	2.3	Univ. Grogingen (NLD) / INRA (FRA)	no
Lc. lactis subsp. *lactis* IL1403	2.0	INRA / Genoscope (FRA)	yes
Leuconostoc citreum KM20	2.0	KRIBB (KOR)	no
Ln. mesenteroides ATCC 8293	2.0	JGI-LABGC (USA)	yes

Lactic acid bacteria continued:			
Pediococcus pentosaceus ATCC 25745	2.0	JGI-LABGC (USA)	no
Streptococcus thermophilis LMD-9	1.8	JGI-LABGC (USA)	no
S. thermophilis LMG18311	1.9	Univ. Catholique de Louvain (BEL)	yes
S. thermophilis CNRZ1066	1.8	INRA (FRA)	no
Other bacteria:			
Bifidobacterium longum NCC2705	2.3	Nestlé (CHE)	yes
B. longum DJ010A	2.1	JGI-LABGC (USA)	yes
B. breve NCIMB8807	2.4	Univeristy College, Cork (IRL)	no
B. linens ATCC9174	3.0	JGI-LABGC (USA)	yes
Propionibacterium freundenreichii ATCC 6207	2.6	DSM Food Specialties (NLD)	no

[1]Public access to nucleotide sequence data provided (effective June 1, 2005)
Abbreviations: JGI-LABGC, Department of Energy Joint Genome Institute and Lactic Acid Bacterial Genomics Consortium; KRIBB, Korea Research Institute of Bioscience and Biotechnology; NR, not reported.

http://wineserver.ucdavis.edu/people/Faculty/mills/LABGC/lab.htm
The value of genome sequence information for dairy-related bacteria to industry and academia cannot be overstated. Access to such comprehensive information databases has endowed researchers with unprecedented power to elucidate mechanisms by which LAB have evolved in, interact with, and respond to, milk and cheese environments. While the fundamental and applied payoffs of genomic research on species that are relevant to the dairy industry are certain to be numerous, it is important to note that sequence acquisition and annotation are only the first steps in genomics research. The physiological role and regulation of many deduced genes must yet be confirmed or identified, and this task could span decades. Nonetheless, global access to LAB genome sequence data and genome mining tools (e.g., microarrays and proteomics) are already driving a number of exciting research advances in dairy microbiology. Examples include studies to elucidate basic mechanisms for global gene regulation and metabolic engineering (*47-49*), cellular adaptation to stress or processing conditions (*50-52*), intestinal growth and probiotic activity (*53,54*), physiological differences among species and strains (*55-57*), and discovery of novel antimicrobials (*58*). Not surprisingly, these and other studies have also provided valuable insight on molecular processes that influence cheese flavor development.

The rapid compilation of LAB genome sequence information in recent years has also accelerated research progress on molecular processes that influence cheese flavor development. As was outlined above, proteolysis and amino acid catabolism are important mechanisms by which LAB contribute to flavor. Genome sequence data and bioinformatics have provided a rapid means to identify enzymes, metabolic pathways, and likely end-products associated with these and other reactions in the cell (*55,56,59*). More importantly, information gleaned from these analyses provides a framework for detailed functional genetics and proteomics-based investigations of cell physiology and its consequences on cheese ripening (*57,60,61*).

Genome-based cheese flavor research by our group is, at present, centered on physiology and enzymology of *Lb. helveticus* CNRZ 32 (fig. 2), a commercial strain that reduces bitterness and intensifies flavor development (*62-64*). We recently assembled a draft (4-fold coverage) genome sequence for CNRZ 32, and have used that sequence to derive an overview of the bacterium's metabolic capabilities in regard to proteolysis, amino acid catabolism, and other flavor-related attributes. As shown in Table II, an immediate outcome of that effort was the identification of several new genes predicted to encode enzymes that would contribute to casein hydrolysis in cheese. Because of the intense and successful research effort Dr. Steele's group had previously afforded this topic (*26*), discovery of more than 20 new genes for proteolytic enzymes in CNRZ 32 was unexpected, and serves to illustrate the capacity of genomics to dramatically accelerate microbiological research. Significantly, functional analysis of three newly discovered endopeptidases, which were of particular interest because

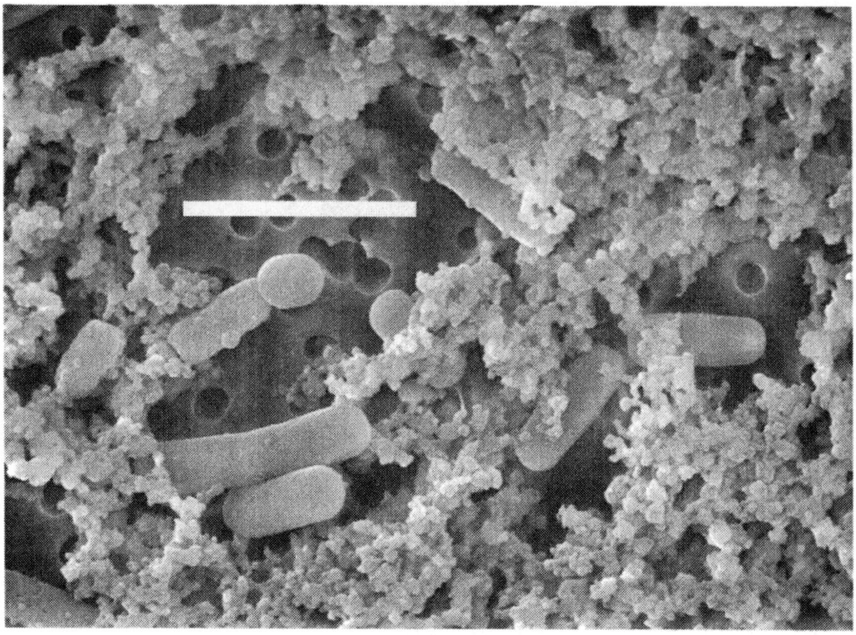

Figure 2. Scanning electron micrograph of Lactobacillus helveticus CNRZ 32. Bar represents 3 μm.

endopeptidases have a central role in hydrolysis of bitter peptides (*65*), showed they likely are critical components of the debittering activity found in this strain (*43*).

Functional genomics is also being used to investigate amino acid catabolism in *Lb. helveticus* CNRZ32. In cheese, the conversion of amino acids into volatile cheese flavor compounds may occur directly or by interaction between starter, adjunct and NSLAB (*66*). The basis for culture interactions is not fully understood, but all LAB are auxotrophic for one or more amino acids, and it has been demonstrated that the primary mechanism for amino acid breakdown by LAB involves the reversible action of enzymes involved in anabolic pathways (*28*). Thus, the interplay that occurs between LAB in amino acid catabolism may reflect the nature of amino acid auxotrophies among different bacteria in cheese. Analysis of the amino acid requirements of *Lb. helveticus* CNRZ 32 by single amino acid omission in a chemically defined medium indicated this bacterium is auxotrophic for 14 amino acids, including Met and all aromatic and branched-chain amino acids (*42*). Reconstruction of amino acid biosynthetic pathways from the *Lb. helveticus* CNRZ 32 genome sequence showed generally good agreement between genome predictions and phenotype, and revealed amino acid

Table II. Genes Discovered in Draft *Lactobacillus helveticus* CNRZ 32 Genome Sequence which are Predicted to Encode Enzymes Involved in Casein Hydrolysis

Proteinases:

prtH2 plus additional proteases

Endopeptidases:

*pepE2, pepF, pepO3,*plus 2 glycoprotein endoproteinases

Aminopeptidases:

pepC2 plus 7 additional aminopeptidases

Di- and Tri-peptidases:

pepD2, pepD3, pepD4, pepQ2, pepT1 and pepT2

Table III. Genetic Basis for Amino Acid Prototrophy or Autotrophy in *Lb. helveticus* CNRZ 32 as Predicted from a 4X Draft Genome Sequence

Amino Acid	*Predicted pathway Status*	*Essential*
Ala	complete; from pyruvate	no
Arg	incomplete; most genes absent	yes
Asn/Asp	incomplete; pseudogene	asn or asp
Glu	incomplete; most genes absent	yes
Gln	complete; from Glu	no
Cys	complete; from Ser	no
Met	incomplete; pseudogene	yes
Lys	incomplete; 1 gene absent	yes
Ile, Leu, Val	incomplete; most genes absent	yes
Phe, Tyr, Pro	incomplete; most genes absent	yes
His	incomplete; all genes absent	yes
Pro	incomplete; all genes absent	yes
Ser	complete; from Gly, glycerate or pyruvate	no
Thr	incomplete; all genes absent	yes
Gly	complete; from Ser	no

auxotrophy in this species was primarily due to complete gene loss rather than point mutations or minor genetic lesions (Table III). Nonetheless, phenotypic data indicate that "remnant" enzymes from truncated pathways, such as the aromatic and branched-chain aminotransferases, contribute to amino acid breakdown by *Lb. helveticus* CNRZ 32 (*67,68*). Because primary sequences of many enzymes involved in amino acid anabolic and catabolic reactions are relatively well-conserved, growing access to genome sequence information is expected to dramatically enhance our ability to predict and test pathways for amino acid catabolism in cheese.

In summary, many of the mechanisms by which LAB affect cheese flavor have been identified, but important details regarding the pathways and enzymes involved in these reactions remain unknown. However, the rapid output of LAB genome sequence information seen in recent years, combined with technological advances in molecular biology tools, computers and software platforms has provided researchers with unprecedented opportunity to address these and other important questions related to LAB genetics and physiology. The knowledge base that will emerge from these efforts will transform the dairy cultures industry and contribute innovation and economic vitality to the cheese and fermented milk industry.

References

1. Axelsson, L. In *Lactic Acid Bacteria: Microbiology and Functional Aspects;* Salminen S.; von Wright, A., Eds.; Marcel Dekker, Inc., New York, NY, 1998; 2nd ed., pp. 1-72.
2. Broadbent, J. R. In *Applied Dairy Microbiology;* Steele, J. L.; Marth, E. H., Eds.; Marcel Dekker, Inc., NY, 2001; 2nd ed., pp. 243-299.
3. Fox, P. F.; Law, J.; McSweeney, P. L. H.; Wallace, J. In *Cheese: Chemistry, Physics and Microbiology*; Fox, P. F., Ed.; Chapman and Hall, London, England, 1993; Vol. 1, 2nd ed., pp. 389-438.
4. Peterson, S.D.; Marshall, R.T. *J. Dairy Sci.* **1990**, *73*, 1395-1410.
5. Fryer, T. F. *Dairy Sci. Abstr.* **1969**, *31*, 471-490.
6. Beresford, T. P.; Fitzsimons, N. A.; Brennan, N. L.; Cogan, T. M. *Int. Dairy J.* **2001**, *11*, 259-274.
7. Broadbent, J. R.; Houck, K.; Johnson, M. E.; Oberg, C. J. *J. Dairy Sci.* **2003**, *86*, 2773-2782.
8. Broome, M. C.;Krause, D. A.; Hickey, M. W. *Aust. J. Dairy Technol.* **1990**, *45*, 60-66.
9. Crow, V.; Curry, B.; Hayes, M. *Int. Dairy J.* **2001**, *11*, 275-283.
10. Sherwood, I. R. *J. Dairy Res.* **1939**, *10*, 449-454.
11. Law, B. A.; Castanon, M.; Sharpe, M. E. *J. Dairy Res.* **1976**, *43*, 117-125.
12. Reiter, B.; Fryer, T. F.; Pickering, A.; Chapman, H. R.; Lawrence, R. C.; Sharpe, M. E. *J. Dairy Res.* **1967**, *34*, 257-272.

13. Broome, M. C.;Krause, D. A.; Hickey, M. W. *Aust. J. Dairy Technol.* **1990**, *45*, 67-73.
14. Lynch, C. M.; Muir, D. D.; Banks, J. M.; McSweeney, P. L. H.; Fox, P. F. *J. Dairy Sci.* **1999**, *82*, 1618-1628.
15. McSweeney, P. L. H.; Walsh, E. M.; Fox, P. F.; Cogan, T. M.; Durain, F. D.; Castelo-Gonzalez, M. *Irish J. Agricul. Food Res.* **1994**, *33*, 183-192.
16. Swearingen, P. A.; O'Sullivan, D. J.; Wartheson, J. J. *J. Dairy Sci.* **2001**, *84*, 50-59.
17. Johnson, M. E.; Riesterer, B. A.; Olsen, N. F. *J. Dairy Sci.* **1990**, *73*, 1145-1149.
18. Khalid, N. M.; Marth, E. H. *J. Dairy Sci.* **1990**, *73*, 2669-2684.
19. El Soda, M.; Madkor, S. A.; Tong, P. S. *J. Dairy Sci.* **2000**, *83*, 609-619.
20. Fox, P. F.; Wallace, J. M. *Adv. Appl. Microbiol.* **1997**, *45*, 17-85.
21. Marilley, L; Casey, M. G. *Int. J. Food Microbiol.* **2004**, *90*, 139-159.
22. Rattray, F. P.; Fox. P. F. *J. Dairy Sci.* **1999**, *82*, 891-909.
23. Lawrence, R. C.; Gilles, J.; Creamer, L. K. In *Cheese: Chemistry, Physics and Microbiology*; Fox, P. F., Ed.; Chapman and Hall, London, England, 1993; Vol. 2, 2nd ed., pp. 1-38.
24. Olson, N. F. *FEMS Microbiol. Rev.* **1990**, *87*, 131-148.
25. Stepaniak L. *Int. J. Dairy Technol.* **2004**, *57*, 153-171.
26. Christensen, J. E.; Dudley, E. G.; Pederson, J. A.; Steele, J. L. *Antonie van Leeuwenhoek* **1999**, *76*, 217-246.
27. Fox, P. F.; Singh, T. K.; McSweeney, P. L. H. In *Biochemistry of milk products*; Andrews, A.T.; Varley, J., Eds.; Royal Society of Chemistry, Cambridge, England, 1994; pp. 1-31.
28. Yvon, M.; Rijnen, L. *Int. Dairy J.* **2001**, *11*, 185-202.
29. Tanous, C.; Kieronczyk, A.; Helinck, S.; Chambellon, E.; Yvon, M. *Antonie Van Leeuwenhoek* **2002**, *82*, 271-278.
30. Smit, G.; Smit, B. A.; Engels, W. J. M. *FEMS Microbiol. Rev.* **2005**, *in press*.
31. Weimer, B.; Seefeldt, K.; Dias, B. *Antonie Van Leeuwenhoek* **1999**, *76*, 247-261.
32. Avsar, Y. K.; Karagul-Yuceer, Y.; Drake, M. A.; Singh, T. K.; Yoon, Y.; Cadwallader, K. R. *J. Dairy Sci.* **2004**, *87*, 1999-2010.
33. Carunchia Whetstine, M. E.; Cadwallader, K. R.; Drake, M. *J. Agric. Food Chem.* **2005**, *53*, 3126-132.
34. Dunn, H. C.; Lindsay, R. C. *J. Dairy Sci.* **1985**, *68*, 2859-2874.
35. Battistotti, B.; Corradini, C. In *Cheese: Chemistry, Physics and Microbiology*; Fox, P. F., Ed.; Chapman and Hall, London, England, 1993; Vol. 2, 2nd ed., pp. 221-244.
36. Fenster, K. M.; Rankin, S. A.; Steele, J. L. *J. Dairy Sci.* **2003**, *86*, 2818-2825.
37. Liu, S. Q.; Holland, R.; Crow, V. L. *Int. Dairy J.* **2004**, *14*, 923-945.

38. Collins, Y. F.; McSweeney, P. L. H.; Wilkinson, M. G. *Int. Dairy J.* **2003**, *13*, 841-866.
39. Dudley, E. G.; Steele, J. L. *J. Appl. Microbiol.* **2005**, *98*, 14-23.
40. de Vos, W. M. *Antonie van Leeuwenhoek* **1996**, *70*, 223-242.
41. Chen, Y. S.; Christensen, J. E.; Broadbent, J. R.; Steele, J. L. *Appl. Environ. Microbiol.* **2003**, *69*, 1276-1282.
42. Christensen, J. E.; Steele, J. L. *J. Bacteriol.* **2003**, 185, 3297-3306.
43. Sridhar, V. R.; Hughes, J. E.; Welker, D. L.; Broadbent, J. R.; Steele, J. L. *Appl. Environ. Microbiol.* **2005**, *71*, 3025-3032.
44. Davidson B. E.; Kordias, N.; Dobos, M.; Hillier, A. J. *Antonie van Leeuwenhoek* **1996**, *70*, 161-183.
45. Le Bourgeois, P.; Lautier, M.; Ritzenthaler, P. *FEMS Microbiol. Rev.* **1993**, *12*, 109-123.
46. Klaenhammer, T.; Altermann, E.; Arigoni, F.; Bolotin, A.; Breidt, F.; Broadbent, J.; Cano, R.; Chaillou, S.; Deutscher, J.; Gasson, M.; van de Guchte, M.; Guzzo, J.; Hartke, A.; Hawkins, T.; Hols, P.; Hutkins, R.; Kleerebezem, M.; Kok, J.; Kuipers, O.; Lubbers, M.; Maguin, E.; McKay, L.; Mills, D.; Nauta, A.; Overbeek, R.; Pel, H.; Pridmore, D.; Saier, M.; van Sinderen, D.; Sorokin, A.; Steele, J.; O'Sullivan, D.; de Vos, W.; Weimer, B.; Zagorec, M.; Siezen, R. *Antonie van Leeuwenhoek.* **2002**, *82*, 29-58.
47. Andersson, U.; Molenaar, D.; Radstrom, P; de Vos, W. M. *Syst. Appl. Microbiol.* **2005**, *28*, 187-195.
48. Kleerebezem, M.; Hugenholtz, J. *Curr. Opin. Biotechnol.* **2003**, *14*, 232-237.
49. Smid, E. J.; Enckevort, F. J.; Wegkamp, A.; Boekhorst, J.; Molenaar, D.; Hugenholtz, J.; Siezen, R. J.; Teusink, B. *J. Appl. Microbiol.* **2005**, *98*, 1326-1331.
50. Pedersen, M. B.; Iversen, S. L.; Sorensen, K. I.; Johansen, E. *FEMS Microbiol. Rev.* **2005**, *in press*.
51. Redon, E.; Loubiere, P.; Cocaign-Bousquet, M. *J. Bacteriol.* **2005**, *187*, 3589-3592.
52. Xie, Y.; Chou, L. S.; Cutler, A.; Weimer, B. *Appl. Environ. Microbiol.* **2004**, *70*, 6738-6747.
53. Barrangou, R.; Altermann, E.; Hutkins, R.; Cano, R.; Klaenhammer, T. R. *Proc. Natl. Acad. Sci. USA.* **2003**, *100*, 8957-8962.
54. de Vos, W. M.; Bron, P. A.; Kleerebezem, M. *Curr. Opin. Biotechnol.* **2004**, *15*, 86-93.
55. Boekhorst, J.; Siezen, R. J.; Zwahlen, M. C.; Vilanova, D.; Pridmore, R. D.; Mercenier, A.; Kleerebezem, M.; de Vos, W. M.; Brussow, H.; Desiere, F. *Microbiol.* **2004**, *50*, 3601-3611.
56. Bolotin, A.; Quinquis, B.; Renault, P.; Sorokin, A.; Ehrlich, S. D.; Kulakauskas, S.; Lapidus, A.; Goltsman, E.; Mazur, M.; Pusch, G. D.; Fonstein, M.; Overbeek, R.; Kyprides, N.; Purnelle, B.; Prozzi, D.; Ngui,

K.; Masuy, D.; Hancy, F.; Burteau, S.; Boutry, M.; Delcour, J.; Goffeau, A.; Hols, P. *Nat Biotechnol.* **2004**, *22*, 1554-1558.

57. Kok, J.; Buist, G.; Zomer, A. L.; van Hijum, S. A.; Kuipers, O. P. *FEMS Microbiol Rev.* **2005**, *in press.*
58. Nes, I. F.; Johnsborg, O. *Curr. Opin. Biotechnol.* **2004**, *15*, 100-104.
59. van Kranenburg, R.; Kleerebezem, M.; Vlieg, J. V.; Ursing, B. M.; Boekhorst, J.; Smit, B. A.; Ayad, E. H. E.; Smit, G.; Siezen, R. J. *Int. Dairy J.* **2002**, *12*, 111-121.
60. Gagnaire, V.; Piot, M.; Camier, B.; Vissers, J. P.; Jan, G.; Leonil, J. *Int. J. Food Microbiol.* **2004**, *94*, 185-201.
61. Sperandio, B.; Polard, P.; Ehrlich, D. S.; Renault, P.; Guedon, E. *J. Bacteriol.* **2005**, *187*, 3762-3778.
62. Bartels, H.J.; Johnson, M.E.; Olson, N.F. *Milchwissenschaft* **1987**, *42*, 83-88.
63. Bartels, H.J.; Johnson, M.E.; Olson, N.F. *Milchwissenschaft* **1987**, *42*, 139-144.
64. Broadbent, J.; Brennand, C.; Johnson, M.; Steele, J.; Strickland, M.; Weimer, B. *Dairy Ind. Int.* **1997**, *62*, 35-39.
65. Christensen, J. E.; Broadbent, J. R.; Steele, J. L. *Appl. Environ. Microbiol.* **2003**, *69*, 1283-1286.
66. Kieronczyk, A.; Skeie, S.; Langsrud, T.; Yvon, M. *Appl. Environ. Microbiol.* **2003**, *69*, 734-739.
67. Gummalla, S.; Broadbent, J. R. *J. Dairy Sci.* **2001**, *84*, 1011-1019.
68. Gummalla, S.; Broadbent, J. R. *J. Dairy Sci.* **1999**, *82*, 2070-2077.

Chapter 11

Proteolytic Enzymes of Lactic Acid Bacteria and Their Influence on Bitterness in Bacterial-Ripened Cheeses

Jeffery R. Broadbent[1] and James L. Steele[2]

[1]Department of Nutrition and Food Sciences and Western Dairy Center, Utah State University, Logan, UT 84322
[2]Department of Food Science, University of Wisconsin, Madison, WI 53706

Bitterness, a common flavor defect in Cheddar and Gouda cheese, results from the accumulation of hydrophobic bitter peptides to concentrations higher than their taste thresholds. Formation of these peptides is directly related to the activity and specificity of the lactococcal cell envelope protease (lactocepin) and the coagulant, chymosin. Degradation of these peptides is related to the activity and specificity of peptidases derived from the starter and non-starter bacteria present in the ripening cheese. Therefore, by controlling the activity and specificity of the proteolytic enzymes of lactic acid bacteria it should be possible to control bitterness. This paper will focus on the contribution of lactocepin to the formation of bitter peptides and a method for altering lactocepin specifically. Additionally, efforts to characterize the proteolytic enzyme system of an adjunct culture known to reduce bitterness, *Lactobacillus helveticus* CNRZ32, will be discussed.

Proteolysis in bacterial-ripened cheese is a complex and dynamic process that involves endogenous milk enzymes, rennet, and microbial proteinases and peptidases. Hydrolysis of intact caseins (CNs) is catalyzed almost exclusively by rennet and endogenous milk enzymes, while the proteinases and peptidases from lactic acid bacteria (LAB) are principally responsible for the production of water-soluble peptides and free amino acids (*1*,*2*). The concerted action of these enzymes in cheese is influenced by the specificity and activity of the individual enzymes present in the cheese matrix and, in the case of intracellular enzymes, access to an appropriate substrate. Studies in Cheddar- and Dutch-type cheeses suggest that the concerted action of these enzymes is one of the most important biochemical events during maturation (*3*). Proteolysis is also associated with bitterness. However, and this flavor defect is commonly observed in reduced- and full-fat Cheddar cheese (*4-6*).

Bitterness develops when small hydrophobic peptides produced from CNs by the coagulant and starter cell envelope proteinase (lactocepin) accumulate to levels that exceed desirable taste thresholds (*5*,*7*). Although intact CNs are flavorless, each casein contains regions of hydrophobicity which can yield bitter peptides during proteolysis. Bitter peptides produced by rennet and proteinases from starter LAB can, however, be further hydrolyzed to non-bitter peptides and amino acids by intracellular peptidases released as a consequence of autolysis of LAB (*5*,*7*). Nonetheless, strain propensity for autolysis and the individual activity of peptidases may vary considerably (*8*,*9*), and research has demonstrated that lactocepin-derived peptides still accumulate in cheese made with autolytic *Lactococcus* (*Lc.*) *lactis* starter bacteria (*10-12*). It has therefore been our hypothesis that the most effective strategy to control bitterness is to develop starter systems that combine a low propensity for bitter peptide production with highly active debittering peptidases. To test this hypothesis, our group has investigated the contribution of lactocepin specificity to the production of bitterness, and the ability of *Lactobacillus* (*Lb.*) *helveticus* peptidases to hydrolyze starter- and rennet-produced bitter peptides.

Proteolytic Enzymes of Lactic Acid Bacteria

The proteolytic enzyme system of LAB represents a diverse compilation of enzymes the collective function of which is primarily devoted to the procurement of essential amino acids and housekeeping needs (e.g., protein processing and turnover). Although milk is an excellent substrate for many LAB, levels of free amino acids and small peptides in milk only can support 2 to 4 cell generations (at 1% inoculum) (*13*). Once those resources are exhausted, cells must be able to derive additional amino acids from milk proteins, particularly casein.

Detailed analysis of proteolytic systems in several dairy LAB, notably *Lc.*

lactis, *Lb. helveticus* and *Lb. casei,* suggest that enzymes involved in the procurement of essential amino acids can be loosely divided into three major categories (*14-16*): 1) extracellular, cell envelope-associated proteinases (e.g., lactocepin) that degrade caseins or large molecular-weight peptides into medium and small molecular-weight oligopeptides; 2) a number of specialized transport systems to take up free amino acids and oligopeptides from the environment (lactococci, for example, possess a minimum of ten amino acid transport systems, two systems for di- and tri-peptide transport, and an oligopeptide transport system that accommodates peptides of 4 to [at least] 18 amino acids in length); and 3) intracellular peptidases that degrade peptides into smaller oligopeptides (endopeptidases) and free amino acids (exopeptidases).

While the biological role of the proteolytic system is focused simply on cellular growth needs, its activity in fermented milk foods has important practical consequences on the sensory, functional and nutritional attributes of these products. The degree of proteolysis in fermented milks, for example, affects the structure and nutritional value of these products, and peptides and amino acids liberated by this activity also serve as precursors for flavor-generating reactions (*17*). Proteolysis and its secondary reactions have an even more pronounced effect on flavor maturation in and functional properties of cheese (*3,18*), and the relationship between proteolytic enzyme activity and flavor development in bacterial-ripened cheese has been a focal point in dairy research for several decades. Those efforts have determined that primary and secondary proteolysis influences cheese flavor development in at least 3 important ways (3): 1) breakdown of the casein network by rennet and plasmin softens cheese texture, which facilitates the release of flavor compounds during chewing; 2) some of the peptides produced by the action of these enzymes are small enough to influence flavor (although their effect is typically negative; i.e., bitter); and 3) high and medium molecular weight peptides that are produced by rennet and plasmin serve as substrates for microbial proteinases and peptidases, which convert them into low molecular-weight peptides and free amino acids. Peptides from the latter reactions may contribute to bitterness (*5*), and free amino acids can be catabolized into a variety of potent flavor compounds (*19*). In the following sections, we outline the role of LAB proteinases and peptidases in the production and degradation of bitter peptides, and offer biotechnological strategies to control bitterness in cheese.

Lactocepin Specificity and Bitterness

As noted above, the degradation of casein or large molecular-weight peptides by LAB in milk or cheese is initiated by an extracellular, cell envelope-associated proteinase (CEP). Although CEPs from several species of dairy LAB have been characterized at the biochemical or genetic level, the most thoroughly studied enzyme in this group is the *prtP* gene product of *Lc. lactis* (*8,14,16*). In Cheddar cheese, the lactococcal cell envelope-associated proteinase, termed lactocepin, is the most important microbial enzyme for the hydrolysis of high- and medium-molecular weight peptides produced by rennet or plasmin. Lactocepin is a 180-190 kDa enzyme that belongs to the subtilisin family of serine proteases. Although lactocepins from different lactococcal strains exhibit more than 98% amino acid sequence identity, purified enzymes may differ in their relative affinity and specificity for individual caseins (*8*). Biochemical studies originally identified two major types of lactococcal proteinases on the basis of their specificity toward α_{s1}- and β-caseins; P_{III}-type enzymes degrade both caseins, while P_I enzymes preferentially degrade β-casein and show little activity toward α_{s1}-casein. More detailed studies showed that lactococcal CEPs may also differ in their specificity toward individual caseins, and DNA sequence analysis of *prtP* genes showed that differences in specificity were linked to one or more amino acid substitutions in the enzyme's substrate-binding regions (*8*). As a result of these observations, eight distinct specificities, designated a-h, have replaced the P_I/P_{III} designation for the classification of lactocepin specificity (*2,20*) (Table I).

Table I. Classification of the *Lactococcus lactis* Cell Envelope-Associated Proteinase (CEP) from Deduced Amino Acid Sequences for Substrate Binding Regions

Prototype strain	*Residues in substrate-binding regions*[1]									*CEP*
	131	*138*	*142*	*144*	*166*	*177*	*747*	*748*	*763*	
SK11	Ser	Lys	Ala	Val	Asn	Leu	Arg	Lys	Asn	a
AM2	Thr	Thr	Ala	Leu	Asp	Leu	Arg	Lys	Asn	b
E8	Thr	Thr	Ala	Leu	Asp	Ile	Arg	Lys	Asn	c
UC317	Thr	Thr	Ala	Leu	Asp	Leu	Arg	Lys	His	d
Wg2	Thr	Thr	Ser	Leu	Asp	Leu	Leu	Thr	Asn	e
Z8	Thr	Thr	Ala	Leu	Asp	Leu	Leu	Thr	His	f
HP	Thr	Thr	Asp	Leu	Asp	Ile	Leu	Thr	His	g
S3	Thr	Thr	Ala	Leu	Gly	Leu	Leu	Thr	His	h

[1]Numbering system is based on the primary sequence for the SK11 group a CEP

While the specificity of purified lactocepins toward α_{S1}-, β- and κ-caseins has been well characterized, the specificity of purified enzymes differ from the native (cell-bound) form and it may also be influenced by the pH, salt content and water activity of cheese (*21,22*). Thus, even though the lactocepin plays an important role in cheese proteolysis and flavor development, knowledge of the relationship between lactocepin specificity and cheese quality is largely incomplete. One notable exception, however, involves the contribution of this enzyme to bitter flavor defect, where several studies have implicated lactocepin specificity in the production of bitter peptides in cheese (*10,23,24*). However, efforts to define the role of lactocepin specificity in bitterness were confounded by strain variability in the propensity for autolysis and intracellular peptidase activity.

To overcome these limitations, Broadbent et al. (*25*) investigated peptide accumulation and bitter flavor development in 50% reduced-fat Cheddar cheeses made with isogenic, autolysis-resistant, single-strain *Lc. lactis* starters that produced group a, e, h or no lactocepin. Results from this and related work confirmed that lactocepin specificity is a primary determinant in the propensity of the starter for bitter flavor development, and showed that this trait can be attenuated or enhanced by exchanging the lactocepin gene, *prtP*, for an allele encoding alternative group specificity (*25,26*). The latter observation is particularly significant because it affords a ready biotechnological strategy to reduce the propensity for bitter peptide production among existing commercial strains. For example, since lactococcal *prtP* is commonly encoded by plasmid DNA, one approach is to exchange any plasmid encoding bitter-type lactocepin for a plasmid-coded non-bitter *prtP* allele (*26*). An alternative strategy that was recently demonstrated by our group (*27*) involves genetic conversion of substrate-binding determinants in a bitter (group h) lactocepin to corresponding regions from a non-bitter (group b) enzyme. Characterization of isogenic *prtP* derivatives recovered from this work confirmed that lactocepin specificity in two industrial strains had been altered in a manner that reflected newly acquired substitutions in the enzymes' substrate binding regions.

Debittering Activity of *Lb. helveticus* CNRZ 32

Because bitter peptides are produced by the action of both rennet and lactocepin, effective control over this problem also requires delivery of enzymes that can hydrolyze these peptides into smaller, non-bitter peptides and free amino acids. In cheese, the latter activity is supplied by LAB endopeptidases and exopeptidases. As was noted earlier in this chapter, endopeptidases hydrolyze peptide bonds within peptides, while exopeptidases hydrolyze peptide bonds at the amino- or carboxy-terminus of peptides. To date, there is little compelling evidence for carboxypeptidases in LAB, so it appears that most (and perhaps all) exopeptidases from these organisms cleave peptides at the amino terminal end (*8,14*).

The complexity of the peptidase enzyme system in LAB has been an area of considerable research effort; many of the enzymes have been purified and characterized, and numerous genes encoding these enzymes have been cloned and sequenced (*8*,*14*) (Table II). Exopeptidases which have been isolated and characterized include two general aminopeptidases (PepN and PepC), aminopeptidase A, dipeptidases and tripeptidases. General aminopeptidases display broad specificity in their ability to sequentially remove amino acids from the amino terminus of peptides containing a variety of different amino acids at this position. Aminopeptidase A removes acidic amino acid residues (aspartyl and glutamyl) from the amino terminus of peptides. Tripeptidases cleave a variety of tripeptides into dipeptides and amino acids, while dipeptidases cleave dipeptides into amino acids (*14*).

Table II. Genes Encoding Peptidases in *Lactobacillus helveticus* CNRZ 32

Endopeptidases:

pepE, pepE2, pepF, pepO, pepO2, pepO3, plus 2 glycoprotein endopeptidases

Exopeptidases:

I. Aminopeptidases:

pepC, pepC2, pepN, pepX plus 7 additional uncharacterized aminopeptidases

II. Di- and tri-peptidases:

pepD, pepD2, pepD3, pepD4, pepI, pepQ, pepQ2, pepR, pepT1 and *pepT2*

The unique structure of proline requires specialized enzymes to hydrolyze peptide bonds that involve this amino acid, and several enzymes involved in the hydrolysis of proline-containing peptides have been identified in LAB. These include: proline iminopeptidase (PepI), which removes Pro residues from the amino terminal of peptides; prolidase (PepQ), which cleaves dipeptides that contain Pro at the 2nd position; and X-prolyl dipeptidyl aminopeptidase (PepX), which removes dipeptides from the amino end of peptides that have Pro at the penultimate position. Peptidases that are active on Pro-containing peptides are of particular interest because of the relatively high abundance of Pro in casein.

In the past, the complexity of the LAB peptidase system complicated efforts to characterize the role of individual enzymes in casein hydrolysis and cheese ripening. However, advances in recombinant DNA (rDNA) technology now allow researchers to construct isogenic strain variants that differ only in the activity of a single peptidase. In recent years, this approach has been used to systematically determine the individual and collective contribution of various peptidases to cell growth and cheese properties (*14,28-31*). Dr. Steele's research group, for example, has utilized rDNA technology to dissect the proteolytic enzyme system of *Lb. helveticus* CNRZ 32, a commercial strain that reduces bitterness and intensifies flavor development in cheese (*32-34*). As is shown in Table II, their research led to the identification and characterization of numerous genes encoding endopeptidases (*28,31,35,36*), aminopeptidases (*37,38,39*) and di-tripeptidases (*14,40,41*). Significantly, these studies also produced a broad catalog of CNRZ 32 null mutants, constructed by gene replacement through rDNA methods, that lacked genes for one or more specific endo- or exo-peptidases (e.g., pepE, pepO, pepC, pepN, and pepX).

Access to these mutants allowed Christensen et al. (*42*) to investigate the individual contribution of CNRZ 32 PepE, PepO, PepN, PepC and PepX to the hydrolysis of two casein-derived bitter peptides, α_{S1}-CN (f 1-9) and β-CN (f 193-209), at pH 6.5 and under the pH and NaCl conditions associated with ripening cheese (pH 5.1, 4% NaCl). Unexpectedly, neither endopeptidase (PepE and PepO) nor the general aminopeptidase PepC made significant contributions to the breakdown of α_{S1}-CN (f 1-9) or β-CN (f 193-209). Instead, the general aminopeptidase PepN proved to be the most important enzyme in the initial hydrolysis of β-CN (f 193-209), but evidence also indicated an unidentified endopeptidase, with a specificity for peptide bonds at the C-terminal side of Pro residues, had an essential role in β-CN (f 193-209) hydrolysis (*42*). In follow-up work, Chen et al. (*28*) used a creative selection protocol that included a synthetic N-terminally-blocked peptide substrate (N-acetyl-β-CN-[f 203-209]-*p*-nitro-anilide) in a coupled reaction with PepN, to isolate a gene from CNRZ 32 encoding an endopeptidase (designated PepO2) with post-proline specificity. This enzyme hydrolyzed β-CN (f 193-209) at the Pro_{196}-Val_{197}, Pro_{200}-Val_{201} and Pro_{206}-Ile_{207} positions, and also severed the Pro_5-Ile_6 bond of α_{S1}-CN (f 1-9) (*28*).

More recently, Sridhar et al. (*31*) investigated the activities of three additional endopeptidases (PepE2, PepF and PepO3) the genes to which were identified from a draft quality genome sequence of *Lb. helveticus* CNRZ 32, as well as PepE, PepO and PepO2, against β-CN (f 193-209) and α_{S1}-CN (f 1-9) under simulated cheese-ripening conditions (pH 5.1, 4% NaCl and 10°C). Although PepE2 and PepO did not hydrolyze either peptide, PepO3 proved to be a functional paralog of PepO2 and hydrolyzed both peptides in an identical manner. Interestingly, PepF showed post-prolyl specificity toward the Pro_{204}-

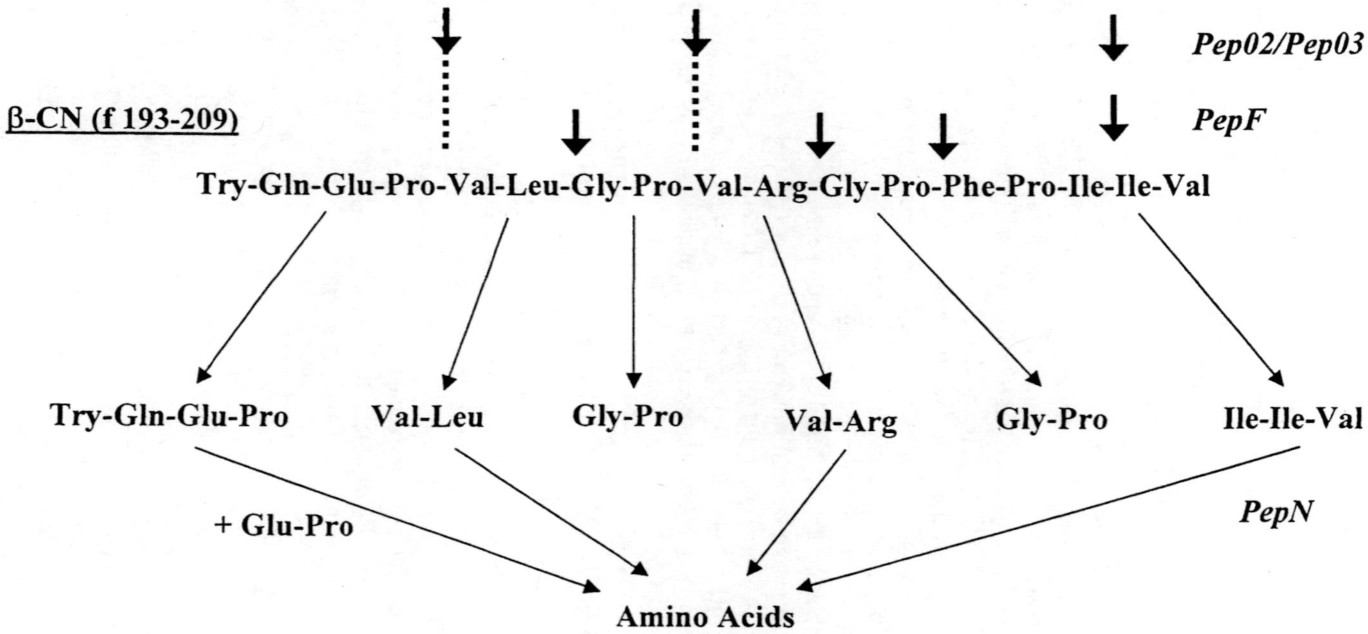

Figure 1. Schematic representation of the breakdown of β-casein (f 193-209) in the cheese matrix by Lactobacillus helveticus CNRZ 32 PepF, PepO2, PepO3 and PepN enzymes as predicted from experiments performed under simulated cheese ripening conditions (pH 5.1, 4% NaCl).

Phe_{205} and Pro_{206}-Ile_{207} positions of β-CN (f 193-209), and also cleaved this peptide at the X-Gly bonds Leu_{198}-Gly_{199} and Arg_{202}-Gly_{203}, but did not hydrolyze α_{S1}-CN (f 1-9). The PepE enzyme hydrolyzed the Lys_3-His_4 and Lys_7-His_8 bonds of α_{S1}-CN (f 1-9) but did not act on β-CN (f 193-209). The same study (*31*) also followed hydrolysis of β-CN (f 193-209) and α_{S1}-CN (f 1-9) by PepE, PepO2 and PepO3, under cheese-ripening conditions, in a defined peptide mix and in the complex peptide background of extracted Cheddar cheese serum. The endopeptidases PepO2 and PepO3 showed the highest activity against β-CN (f 193-209) and α_{S1}-CN (f 1-9) in all of the peptide systems examined (31).

The latter observation is significant because Pro constitutes 16.7% of β-CN and 8.5% of α_{s1}-CN amino acid residues (43), and enzymes that are able to hydrolyze peptide bonds that involve Pro residues are therefore likely to have a central function in the hydrolysis of CN-derived peptides in cheese. Moreover, CN-derived bitter peptides contain relatively large amounts of Pro and the spatial structure that results from Pro incorporation in a peptide is thought to have a direct relationship to bitter flavor perception (*7*). Thus, post-prolyl endopeptidases like PepO2, PepO3 and PepF are probably essential for the debittering activity of *Lb. helveticus* CNRZ 32 because they convert Pro-containing bitter peptides such as β-CN (f 193-209) into smaller peptides that can then be efficiently degraded to free amino acids and non-bitter X-Pro dipeptides by the general aminopeptidase PepN (Fig. 1).

In summary, research by our group on the role of specific microbial enzymes in the production or degradation of bitter peptides in Cheddar cheese has shown that the specificity of the lactococcal cell envelope proteinase is a primary determinant of the propensity of the starter for bitterness, and has identified several *Lb. helveticus* CNRZ 32 peptidases that degrade known bitter peptides in cheese systems. This information has provided a knowledge base for processors to select starter adjunct combinations that can effectively control bitterness in cheese.

References

1. Fox, P. F.; Singh, T. K.; McSweeney, P. L. H. In *Biochemistry of milk products*; Andrews, A.T.; Varley, J., Eds.; Proteolysis in cheese during ripening; Royal Society of Chemistry, Cambridge, England, 1994; pp. 1-31.
2. Stepaniak L. *Int. J. Dairy Technol.* **2004**, *57*, 153-171.
3. Fox, P. F.; Law, J.; McSweeney, P. L. H.; Wallace, J. In *Cheese: Chemistry, Physics and Microbiology*; Fox, P. F., Ed.; Biochemistry of cheese ripening; Chapman and Hall, London, England, 1993; Vol. 1, 2nd ed., pp. 389-438.
4. Jameson, G. W. *Aust. J. Dairy Technol.* **1990**, *37*, 93-98.
5. Lemieux, L.; Simard, R. E. *Lait.* **1991**, *71*, 599-636.

6. Olson, N. F.; Johnson, M. E. *Food Technol.* **1990**, *44*, 93-96.
7. Lemieux, L.; Simard, R. E. *Lait.* **1992**, *72*, 335-382.
8. Kunji, E. R.; Mierau, I.; Hagting, A.; Poolman, B.; Konings, W. N. *Antonie Van Leeuwenhoek.* **1996**, *70*, 187-221.
9. Weimer, B.; Dias, B.; Madhavi, U.; Broadbent, J.; Brennand, C.; Jaegi, J.; Johnson, M.; Milani, F.; Steele, J.; Sisson, D. V. *Lait.* **1997**, *77*, 383-398.
10. Broadbent, J. R.; Strickland, M.; Weimer, B.; Johnson, M. E.; Steele, J. L. *J. Dairy Sci.* **1998**, *81*, 327-337.
11. Kaminogawa, S.; Yan, T. R.; Azuma, N.; Yamauchi, K. *J. Food Sci.* **1986**, *51*, 1253-1256.
12. Singh, T. K.; Fox, P. F.; Healy, A. Int Dairy J. **1994**, *4*, 111-122.
13. Mills, O. E.; and Thomas, T. D. *N. Z. J. Dairy Sci. Technol.* **1981**, *16*, 43-55.
14. Christensen, J. E.; Dudley, E. G.; Pederson, J. A.; Steele, J. L. *Antonie van Leeuwenhoek* **1999**, *76*, 217-246.
15. Mierau, I.; Kunji, E. R. S.; Venema, G.; Kok, J. *Biotechnol. Genet. Eng. Rev.* **1997**, *14*, 279-301.
16. Siezen, R. J. *Antonie Van Leeuwenhoek* **1999**, *76*, 139-155.
17. Oberman, H.; Libudisz, Z. In *Microbiology of fermented foods*; Wood, B. J. B., Ed.; Fermented milks; Blackie Academic & Professional, London, England, 1998; Vol. 1, 2nd ed., pp. 308-350.
18. Rowney, M.; Roupas, P.; Hickey, M. W.; Everett, D. W. *Aust. J. Dairy Technol.* **1999**, *54*, 94-102.
19. Yvon, M.; Rijnen, L. *Int. Dairy J.* **2001**, *11*, 185-202.
20. Exterkate, F. A.; Alting, A. C.; Bruinenberg, P. G. *Appl. Environ. Microbiol.* **1993**, *59*, 3640-3647.
21. Exterkate, F. A.; Alting, A. C. *Syst. Appl. Microbiol.* **1993**, *16*, 1-8.
22. Reid, J. R.; Coolbear, T. *Appl Environ Microbiol.* **1999**, *65*, 2947-2953.
23. Law, J.; Fitzgerald, G. F.; Daly, C.; Fox, P. F.; Farye, N. Y. *J. Dairy Sci.* **1992**, *75*, 1173-1185.
24. Visser, S.; Hup, G.; Exterkate, F. A.; Stadhouders, J. *Neth. Milk Dairy J.* **1983**, *37*, 169-180.
25. Broadbent, J. R.; Barnes, M.; Brennand, C.; Strickland, M.; Houck, K.; Johnson, M. E.; Steele, J. L. *Appl. Environ. Microbiol.* **2002**, *68*, 1778-1785.
26. Pillidge, C. J.; Crow. V. L.; Coolbear, T.; Reid, J. R. *Int. Dairy J.* **2003**, *13*, 345-354.
27. Broadbent, J. R.; Rodriguez, B.; Joseph, P.; Smith, E. A.; Steele, J. L. **2005**, *unpublished.*
28. Chen, Y. S.; Christensen, J. E.; Broadbent, J. R.; Steele, J. L. *Appl. Environ. Microbiol.* **2003**, *69*, 1276-1282.
29. Christensen, J. E.; Steele, J. L. *J. Bacteriol.* **2003**, *185*, 3297-3306.

30. Mierau, I.; Kunji, E. R.; Leenhouts, K. J.; Hellendoorn, M. A.; Haandrikman, A. J.; Poolman, B.; Konings, W. N.; Venema, G.; Kok, J. *J. Bacteriol.* **1996**, *178*, 2794-2803.
31. Sridhar, V. R.; Hughes, J. E.; Welker, D. L.; Broadbent, J. R.; Steele, J. L. *Appl. Environ. Microbiol.* **2005**, *71*, 3025-3032.
32. Bartels, H.J.; Johnson, M.E.; Olson, N.F. *Milchwissenschaft.* **1987**, *42*, 83-88.
33. Bartels, H.J.; Johnson, M.E.; Olson, N.F. *Milchwissenschaft*. **1987**, *42*, 139-144.
34. Broadbent, J.; Brennand, C.; Johnson, M.; Steele, J.; Strickland, M.; Weimer, B. *Dairy Ind. Int.* **1997**, *62*, 35-39.
35. Chen, Y-S.; Steele, J. L. *Appl. Environ. Microbiol.* **1998**, *64*, 3411-3415.
36. Fenster, K. M.; Parkin, K. L.; Steele, J. L. *J. Bacteriol.* **1997**, *179*, 2529-2533.
37. Christensen, J.E.; Lin, D.; Palva, A.; Steele, J. L. *Gene* **1995**, *155*, 89-93.
38. Fernández, L.; Bhowmik, T.; Steele, J. L. *Appl. Environ. Microbiol.* **1994**, *60*, 333-336.
39. Yüksel, G. Ü.; Steele, J. L. *Appl. Microbiol. Biotech.* **1996**, *44*, 766-773.
40. Dudley, E.G.; Husgen, A. C.; He, W.; Steele, J. L. *J. Bacteriol.* **1996**, *178*, 701-704.
41. Shao, W.; Yüksel, G. Ü; Dudley, E. G.; Parkin, K.L.; Steele, J. L. *Appl. Environ. Microbiol.* **1997**, *63*, 3438-3443.
42. Christensen, J. E.; Broadbent, J. R.; Steele, J. L. *Appl. Environ. Microbiol.* **2003**, *69*, 1283-1286.
43. Swaisgood, H. E. In, *Developments in dairy chemistry*; Fox, P. F., Ed.; The chemistry of milk protein; Elsevier, London, England, 1982; Vol. 1, pp. 1-59.

Chapter 12

Volatile Sulfur-Containing Compounds from Methionine Metabolism in Genetically Modified *Lactobacillus helveticus* CNRZ32 Strains

Scott A. Rankin[1], Dattatreya S. Banavara[1], Ed S. Mooberry[2], James L. Steele[1], Jeffery R. Broadbent[3], and Joanne E. Hughes[4]

Departments of [1]Food Sciences and [2]Biochemistry, University of Wisconsin, Madison, WI 53706
Departments of [3]Nutrition and Food Science and [4]Biology, Utah State University, Logan, UT 84322–8700

Cystathionine-ß-lyase (CBL) catalyzes the conversion of methionine to methanethiol, dimethyl disulfide and dimethyl trisulfide. These three compounds are collectively referred to as volatile sulfur compounds (VSC). Two strains of *Lactobacillus helveticus* CNRZ32 were evaluated for VSC production; wild-type and a genetically modified variant that produced elevated levels CBL. Whole cell suspensions and cell free extracts were incubated in buffer (pH 6.0) with methionine substrate; furfuryl alcohol was added as an internal standard. Results demonstrated that the modified strain produced ~2-fold higher levels of VSC, primarily dimethyl disulfide. In parallel, NMR was used to substantiate a lyase-mediated pathway by tracking metabolites of uniformly labeled ^{13}C methionine (17.5mM).

Flavor development in cheese is attributed mainly to the metabolic activities of starter and non-starter lactic acid bacteria. Catabolism of free amino acids yields desirable and undesirable flavor compounds (*1*). One class of flavor compounds, volatile sulfur compounds (VSC), can constitute a major portion of the odor active volatiles in ripened cheeses (*2*). During cheese ripening, microbial proteases, rennet and other enzymes, metabolize casein into smaller peptides. These peptides further yield a pool of amino acids by the action of peptidase enzymes. Methionine and cysteine are believed to be present in cheese at levels up to 0.02%. Methionine has been identified as a main precursor of VSC in certain cheeses (*3*). The catabolism of methionine to VSC's has been studied extensively in a variety of lactic acid bacteria (*4, 5*) and yeasts (*6*). Methionine conversion is believed to take place by at least two major pathways, one initiated by aminotransferases (ATases) and another initiated by lyases such as methionine-γ-lyase and cystathionine-β-lyase (CBL) (*7*). The mechanisms, pathways and intermediate compounds formed are very different for these two enzymes. The ATase pathway has been shown to play a major role in converting methionine to VSC's in lactococcal strains (*8*). Lyases and ATases share a common co-factor, pyridoxal phosphate. However, VSC generation by the ATase-dependent pathway is also contingent on the presence of α-keto acids. In cheese, multiple pathways of VSC generation might exist based on factors such as bacterial strain, pH, and substrate/co-factor/enzyme availability. In this study, *Lactobacillus helveticus* CNRZ32 modified to overexpress the lyase enzyme CBL was studied for its ability to generate increased amounts of VSC's.

Experimental

Chemicals

The compounds L-methionine, α-keto-4-(methylthio)-butanoic acid (KMBA), 2-hydroxy-4-(methylthio)-butanoic acid (HMBA), 2-ketobutyric acid (KBA), NADH, pyridoxal phosphate, α-ketoglutarate (α-KG), dimethyl disulfide (DMDS), dimethyl trisulfide (DMTS), methional and furfuryl alcohol were purchased from Sigma-Aldrich (St Louis, MO). (U)^{13}C Methionine was obtained from Cambridge Isotope Laboratories (Andover, MA).

CBL Overexpression

Recent analysis of a draft quality (4X) genome sequence for *Lb. helveticus* CNRZ32 identified a gene cluster encoding 3 enzymes involved in Cys biosynthesis: O-acetylserine lyase (*cysK*), CBL, and serine acetyltransferase (*cysE*). To further characterize the role of CBL in VSC generation, the *Lb. helveticus* CBL gene was cloned into the expression vector pTRK687 (*9*) then transformed by electroporation into *Lb. helveticus* CNRZ32 as described previously (*10*) . Two independent clones containing the recombinant plasmid (pTRK687:*cbl*) were collected, and the integrity of the plasmid construct was confirmed by DNA sequence analysis. *Lb. helveticus* CNRZ32 wild type cells were also transformed with unmodified pTRK687 and designated LH32. Stock cultures of LH32 and each of the genetically modified variants, designated LH32*cbl.1* and LH32*cbl.2*, were stored at -80°C in nonfat dry milk with 11% glycerol until needed.

Organisms and Cell growth

Working cultures of *Lb. helveticus* LH32, LH32*cbl.1* and LH32*cbl.2* were prepared from frozen stocks by transfer into in 5.5% MRS broth media (Difco, Detroit, MI) that contained chloramphenicol (5μg/ml) with incubation at 42±1°C for 20 hours. Cells were grown in triplicate (45ml media, 90-100μl inocula at comparable OD_{600}). Growth of bacteria was monitored hourly by OD_{600} and pH of media.

Cell harvest and preparation

Cells were harvested in late log phase to early stationary phase (at comparable growth periods), washed twice with potassium phosphate monobasic-disodium phosphate buffer solution (pH 6.0) and pelleted by centrifugation (6,500 × g, 15 min). The pH of the remaining cell free media was noted. The pelleted cells were washed twice by vortexing and centrifuging with a 0.05 M potassium phosphate buffer (adjusted to pH 6.0 with 0.1 N NaOH) at 4°C. Buffer (24 ml) was added to the washed cell pellet with 5 ml of methionine (100 mM) substrate, 100 μl of pyridoxal phosphate (1 mM) co-factor, and 150 μl of furfuryl alcohol (10 μl/ml, internal standard) in a 40 ml glass vial fitted with teflon lined septum cap. After vortexing, 2 ml were removed and analyzed for cell concentration spectrophotometrically at 600 nm. The final OD_{600} of the cell was adjusted to 2.0. Two negative controls were prepared for each sample, one by adding substrate and cofactor to the cells autoclaved for 20 min and the other without any substrate with the native cell suspension. Each of the experimental treatments was prepared in duplicate with two controls.

Cell extracts

Cell extracts were prepared by sonication (Branson 1510, Danbury, CT) of 20 ml cell suspensions under ice at 4°C. A preliminary analysis of cell lysis showed that maximum extracellular protein was obtained in cell free extracts under sonication for 25 min with the sonicator used in the study. Cell debris was removed by centrifugation (14,000 × g, 30 min, 4°C) to yield a crude cell free extract. Protein analysis of the samples was done using a BCA protein assay kit (Pierce Chemicals, Rockford, IL).

SPME-GC-MS

The formation of methionine-derived volatiles was studied using solid phase microextraction (SMPE) with GC-MS analysis. Approximately 30 min prior to each assay, samples incubated for each designated time period were removed, shaken well and allowed to equilibrate for 30 min. in a multiblok heater at 30°C. A SPME fiber coated with carboxen/PDMS (85 µm) was used for adsorbing the volatiles by placing it in the headspace of the equilibrated (static headspace) sample for 10 min. The volatiles absorbed onto the fiber were desorbed into the GC inlet at 280°C. The fiber was held in the injection port over the complete program cycle (17 min) to ensure complete desorption. The GC temperature program was as follows: 35°C initial temp, 2 min hold time, 5°C per min to 70°C, 20°C per min to 200°C, hold for 5min. The quantitative assessments of VSC were done in the selected ion monitoring mode (*m/z* 48 for MTL, *m/z* 94 for DMDS and *m/z* 126 for DMTS). Total mM of VSC was calculated using standard curves. The variation in adsorption efficiencies was corrected by the use of as internal standard (furfuryl alcohol at a concentration noted above).

^{13}C Nuclear Magnetic Resonance

Nuclear magnetic resonance (NMR) has been used in the field of biochemistry and medicine for varied purposes. Nuclei with odd number of protons and neutrons (e.g. ^{1}H, ^{13}C , ^{15}N, ^{19}F, ^{31}P) exhibit a mechanical spin phenomenon. When placed in a static magnetic field, such atoms are driven to a spin balanced state known as polarization. An exciting radiofrequency applied in definite pulses causes perturbation (resonance) and relaxation of the spin balanced state and magnetic resonance signal. Different nuclei resonate at different frequencies and NMR signal of a given nuclei is referred to as chemical shift. The scale of chemical shifts varies for different nuclei. Chemical shifts of nuclei attached to different atoms also vary within the scale. Analysis of molecular transitions and changes can be tracked by analyzing NMR spectra of samples over time. The use of ^{13}C enriched molecules increases the specificity and sensitivity of the method. In this study, (U)^{13}C Met was used as the substrate in whole and cell free extracts. All ^{13}C NMR spectra were obtained with a Bruker model DMX400 wide-bore NMR spectrometer operating at a carbon NMR frequency of 100.6 MHz with a broadband 5-mm-diameter NMR probe at a temperature of 30°C. The ^{13}C NMR spectra were obtained with power-gated proton decoupling as directed by a Bruker pulse program zgpg30 by using the following parameters: ^{13}C spectral window, 225 ppm; 90-degree pulse width, 10 ms; 1 s relaxation delay; 2048 scans per spectrum. The catabolic process was monitored at 2 hr intervals over time.

Results and Discussion

Bacterial Growth

Growth of both CBL expression clones significantly differed from the wild-type strain relative to the onset of exponential growth phase. Fig. 1 represents growth curves for the bacteria as determined by pH and OD_{600}. For analyzing VSC production capability, bacteria were harvested at comparable growth periods (18 hr for CBL expression clones, 16 hr for wild-type).

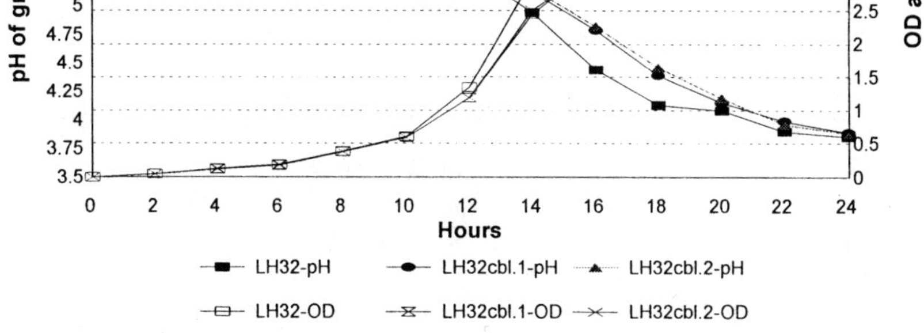

*Figure 1. Growth curve for LH32, LH32*cbl.1 *and LH32*cbl.2 *Strains.*

Production of VSC

In the control samples (autoclaved cells with Met and whole cells without added Met), there were no detectable levels of VSC. There were no differences in VSC concentration trends between studies done using whole cells or cell free extracts. Three VSC's, methanethiol (MTL), dimethyl disulfide (DMDS) and dimethyl trisulfide (DMTS) were found in the headspace of whole cell media of the GM and native treatments. The concentrations of MTL and DMTS found in all the samples were relatively low as compared to DMDS. This is probably due to the fact that MTL readily converts to DMDS in aqueous media via the formation of free radicals especially in presence of oxygen; DMDS is relatively more stable. Among the VSC's found in incubated samples, DMTS formation requires three molecules of MTL while DMDS involves two. The conversion of MTL into DMDS and DMTS may also vary with water activity and pH. Previous research has reported total VSC's as a measure of methionine conversion since total VSC concentration is thought to be related by reactions involving the same pool of MTL. The mechanism of interconversion of MTL to DMDS and DMTS is given in the series of reactions below.

$$CH_3SH \longleftrightarrow CH_3S^- + H^+$$

$$CH_3S^- \longleftrightarrow CH_3S^* + e^-$$

$$CH_3S^* + CH_3S^* \longleftrightarrow CH_3S\text{-}SCH_3 \text{ (predom.)}$$

$$CH_3S\text{-}SCH_3 + e\text{-} \longleftrightarrow CH_3S\text{-}S^- + CH_3^+$$

$$CH_3S\text{-}S^- + CH_3S^* \longleftrightarrow CH_3S\text{-}S\text{-}SCH_3$$

Production of VSC's by LH32, LH32*cbl.1* and LH32*cbl.2* was repetitive under like conditions. Both CBL clones (LH32*cbl.1* and LH32*cbl.2*) produced VSC's at higher concentrations (at least 2 times) as compared to the native strain, LH32 (Fig. 2). VSC production had no correlation to cell autolysis.

Because of the reactive nature of S-compounds and their susceptibility to interconversion as influenced by such factors as redox potential, pH, etc., the ratio of VSC's may vary while the total pool remains constant. However, in the controlled conditions of this study, there were no significant differences in the relative concentrations of MTL, DMDS and DMTS, but, as was predicted from

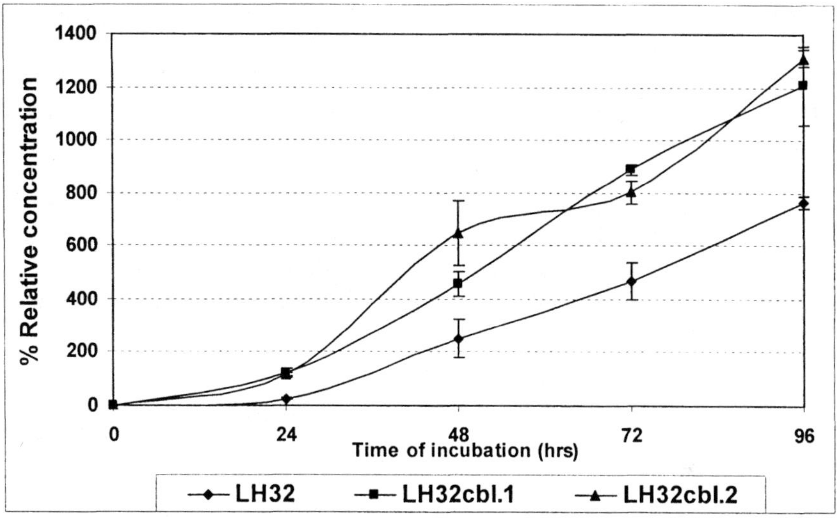

Figure 2. Production of VSC by whole cells of Lb. helveticus *LH32, LH32*cbl.1 *and LH32*cbl.2. *(Error bars represent one standard deviation, n = 2).*

the relative stabilities, DMDS was the predominant VSC by a factor of approximately ten.

^{13}C Nuclear Magnetic Resonance

Gao et al (*8*) demonstrated that *L. lactis* strains follow an ATase mediated pathway to produce VSC from methionine. The presence of α-ketoglutarate (α-KG) is found to be important for this reaction. In this study with *Lb. helveticus*, both wild-type and CBL expression mutants evaluated with NMR analysis did not show any ATase-related pathway intermediates (KMBA or HMBA, Fig. 3). The chemical shifts for DMDS and DMTS were observed with no shifts for KMBA or HMBA.

Lyases are known to hydrolyze methionine yielding MTL, ammonia and KBA. In this study, the chemical shifts for KBA were also not found. There are several reasons for this observation. First, methionine catabolism was slower in *Lb. helveticus* as compared to *L. lactis* studies making the limit of detection a possible issue. Also, once methionine is catabolised by ATase to KMBA or HMBA, conversion to VSC's proceeds slowly, either by redox-driven chemical degradation or via low-activity enzymes and hence KMBA or HMBA may accumulate in the system where they can be more readily detected. Though there was much higher VSC production with the CBL overexpression variants, chemical shifts of ATase-derived intermediates were not observed. Addition of α-KG to cell suspensions, decreased VSC production significantly. It is possible that VSC production predominantly proceeds via a lyase pathway and upon addition of α-KG, the substrate is utilized by both ATase and lyase increasing the intermediates and decreasing VSC. Also, NMR studies did not show any chemical shifts for KMBA, HMBA or KBA under these conditions for lactobacilli even with the addition of α-KG. However, in NMR studies with *L. lactis,* KMBA was previously observed (*8*) as well as denoted in our repeated study with *L. lactis* (Fig. 3).

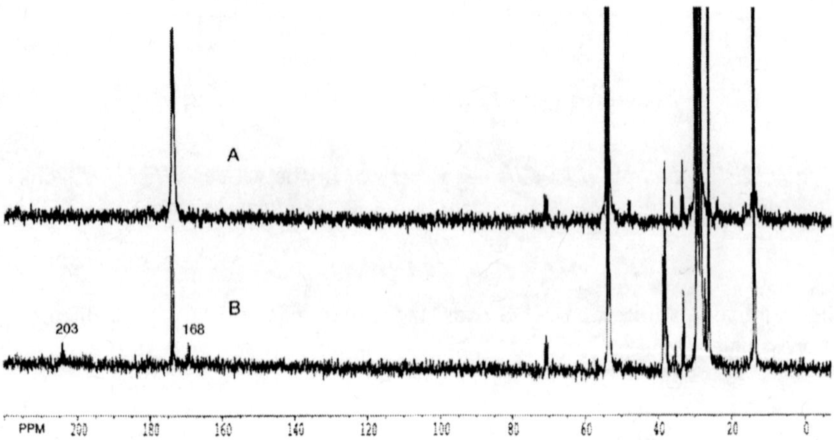

Figure 3. NMR signal of ^{13}C *methionine metabolized by* Lb. helveticus *CNRZ32 (A) and* Lc. lactis *ssp.* cremoris *(B). In B, peaks at 168 and 203 ppm denote the first and second carbon nuclei, respectfully, of KMBA absent in the lactobacillus signal (A).*

References

1. Curtin, A.C.; McSweeney P.L.H. *J. Dairy. Res.* **2003**, *70*, 249-252.
2. Singh, T.K.; Drake, M.A.; Cadwallader, K.R. *Comp. Rev. in Food Sci. and Food Safety* **2003**, *2*, 139-162.
3. Seefeldt, K.E.; Weimer, B.C.. *J.Dairy Sci.* **2000**, *83*, 2740-2746.
4. Helinck, S.; Le Bars, D.; Moreau, D.; Yvon, M. *Appl. Environ. Microbiol.* **2004**, *70(7)*, 3855-3861.

5. Pripis-Nicolau, L.; De Revel, G ; Bertrand, A.; Lonvaud-Funel, A. *J. Appl. Microbiol.* **2004**, *96(5)*, 1176-1184.
6. Dias. B.; Weimer, B. *Appl. Environ. Microbiol* **1998**, *64(9)*, 3320-3326.
7. Tamman. J.D.; William, A.G.; Nobel, J.; Lloyd, D. *Lett .Appl. Microbiol.* **2000**, *30*, 370.
8. Gao. S.; Mooberry, E.S.; Steele, J.L. *Appl. Environ. Microbiol.* **1998**, *64*, 4670-4675.
9. Sturino, J.M.; Klaenhammer, T.R. *Appl Environ Microbiol* **2004**, *70*, 1735-1743.
10. Bhowmik, T.; Steele, J.L. *J. Gen Microbiol.* **1993**, *139*, 1433-1439.

Processing and Storage Effects

Chapter 13

The Flavor and Flavor Stability of Skim and Whole Milk Powders

Mary E. Carunchia Whetstine and MaryAnne Drake

Department of Food Science, Southeast Dairy Food Research Center, North Carolina State University, Campus Box 7624, Raleigh, NC 27695

Skim and whole milk powder (SMP, WMP) are widely used as food ingredients and for direct consumption. Since milk powders are stored prior to use, flavor stability and changes in flavor profiles during storage can impact quality and salability. The objectives of this study were to characterize flavor changes in SMP and WMP throughout 36 months at 21°C using sensory and instrumental methods as well as review sources of flavor formation in milk powders. WMP off-flavor formation occurs as quickly as 3-6 months and is primarily a function of lipid oxidation. SMP flavor is much more variable and some powders develop off-flavors immediately, while others are stable throughout storage. Because off-flavors from milk powders can carry through into product applications, it is important to understand when and how flavor develops to maximize the potential usage of these functional and important ingredients.

Milk powder is an important ingredient in the U.S. for both direct consumption and as an ingredient in foods (*1*). Skim milk powder (SMP) is the most common dry milk product manufactured in the U.S., followed by whole milk powder (WMP) (*2*). In 2003 (most recent statistic available), there were 641 metric tons of SMP and 19 metric tons of WMP produced in the US (*2*). Providing a high quality, consistent product is very important. The dairy industry has long recognized that sensory quality is one of the most critical and important aspects of sales and marketing (*3, 4*).

Milk powder production involves the following steps: receiving, refrigerated storage, standardization, heat treatment, evaporation, homogenization, drying, and packaging (*5, 6*). Additionally, during the production of SMP, the fluid milk must be skimmed to a fat content of not more than 0.1 % (*5*). At all of these steps, there is flavor formation potential (both characteristic- and off-flavors). Milk that is to be dried into milk powder must be of high quality in order to ensure quality and shelf-life in the final product (*5*). More details about flavor formation are discussed in this paper.

There are three general types of SMP produced: low heat SMP (not over 71 °C for 2 minutes), medium heat milk powder (71-79 °C for 20 minutes) and high heat milk powder (88 °C for 30 minutes) (*2*). The heat treatment affects both the physical and chemical properties of the powders due to the differing degree of protein denaturation. Functionality attributes of milk powders include water absorption and binding, foaming, emulsification, solubility, viscosity, gelation, and heat stability (*7*). The sensory properties are also different in these types of products. Increased heating leads to more protein denaturation, and this can liberate sulphydryl groups, giving medium and high heat SMP more cooked/sulfur flavor (*8,9*). WMP is processed using a higher temperature for pasteurization (88-95 °C for 15-30 sec) and is also heated more severely during drying than low heat SMP (*5, 10*). By increasing the temperature, there is more protein denaturation and consequently increased liberation of free sulphydryl groups (*10*). These compounds act as antioxidants, increasing the shelf-life of WMP (*10*). The increased free sulphydryl groups also contribute to cooked flavor. Fresh low heat SMP and WMP should ideally exhibit a mild and bland flavor reminiscent of fluid skim and whole milk, respectively (*3*). Recent research has demonstrated that it certainly is feasible for these products to have flavor profiles similar to their fluid counterparts (*9*).

There are many different sources of flavor and flavor variability in SMP and WMP that will be discussed in this chapter. Besides the flavor formation potential during processing, the quality of fluid milk and storage time and conditions of the milk powders also play a role in flavor. Though it is understood that stored milk powder will not be exactly the same, chemically and physically, as fresh milk powder, customers of these products do expect that the differences will not be large enough that their consumers find the milk powder unacceptable during storage within the projected shelf life (*11*). The objectives of this paper are to review sources of flavor formation in both fresh and stored milk powders as well as systematically study SMP and WMP flavor during storage.

Analysis Techniques

Sensory Techniques

Sensory analysis is a compilation of different tools or tests that can be used for subjective or objective evaluation of food sensory properties. Selection of the appropriate tool or test for a specific objective is required to obtain appropriate and optimal results. Early sensory analysis in the dairy industry was conducted using quality judging and grading. These techniques were established by the federal government in the early 1900's (*12*). These tests are still used today in the industry and are used to assess overall quality based on previously defined defects (*3*). Generally 1-2 expert graders are used and evaluations are not replicated. Because only 1-2 experts are used, and the grading is done by scoring presence/absence of predetermined defects, it is not possible to statistically analyze these results. Additionally, scores for products of a similar quality may be judged to be the same, when in fact, they have vastly different flavor profiles (*13, 14*). Though some research is published using grading defects, this type of analysis leads to data misinterpretation that may have significant impacts on research interpretations and customer or consumer preferences.

There are two main types of sensory analysis techniques: affective and analytical (*15, 16*). Tests that utilize consumers and their perceptions of acceptability are called affective tests. These techniques are important to the food industry because they help explain the role that flavor, texture, and appearance play to consumer acceptability. It is important to keep in mind that these types of techniques can only measure what untrained consumers think and there is much variability from person to person. To increase the value of the information obtained from affective tests, a large number of consumers need to be used (>50) (*15*).

Sensory tests using screened or trained panelists are analytical tests. These include discriminatory tests (difference and threshold) and descriptive sensory analysis, the most powerful tool in the sensory arsenal. When using descriptive sensory analysis, it is important to keep in mind that the panel operates as a group or instrument, and the individual panelists are components of the sensory instrument. Once trained, a descriptive panel operates as a single instrument, and data must be replicated (*15*). This technique is ideal for both identifying flavors in a product as well as discriminating between products (*17*). Descriptive sensory analysis can also be used in conjunction with instrumental analysis to garner a more complete picture of flavor.

There are different types of descriptive sensory analysis, including the Flavor Profile Method, Quantitative Descriptive Analysis (QDA), and the Spectrum technique™, as well as hybrids of the above techniques. These approaches have been fully reviewed elsewhere (*17, 18*). Descriptive sensory analysis is a technique that requires panelists to be familiar with the scaling methods and sensory language (*16, 18*) and it is imperative to maintain constant training to minimize variability in the analyses. Descriptive analysis does not require expensive instrumentation, but time and training by an experienced panel leader is required.

When using descriptive analysis, a sensory language (lexicon) must be developed. Table I lists the dried dairy ingredient lexicon utilized to evaluate SMP and WMP. A lexicon that has well defined references can be used across different panel sites and leaders (*19, 20*) and can be used to link instrumental and sensory terms to identify specific flavors (*4, 18*).

In this study, a trained sensory panel (n=7) evaluated the flavor attributes of the reconstituted WMP and SMP using a previously published lexicon for dried dairy ingredients (*1, 9, 21*). Two WMP and two SMP (commercially packaged in 25 kg 2 ply paper bags with liners) were collected from manufacturers within 48 h of production. Powders were stored at 21 °C in the dark and sampled after 3, 6, 9, 12, 18, and 24 months for WMP and 3, 6, 9, 12, 18, 24, 30, and 36 months for SMP. For sensory and instrumental analyses, WMP were rehydrated using the formula: 1000/100-dry % fat content of WMP = g of WMP in 90 g water (*22*) and SMP were rehydrated to 10% solids with deodorized water (prepared by boiling 4 L of distilled water until its volume was decreased by one-third) and blended with an electric hand-held mixer. Samples were reconstituted 24 h prior to evaluation.

Panelists each received 100 h training on aroma and flavor evaluation of dried dairy ingredients, including both SMP and WMP. Flavor and taste intensities were scaled using a 15-point intensity scale using the Spectrum™ descriptive analysis method (*16, 18*). Samples were evaluated in duplicate at 12 °C on separate occasions in 2 oz lidded plastic cups with three digit random codes.

Instrumental Techniques

Descriptive sensory analysis is performed to determine the flavor profile of a food. This is a very powerful technique, but sometimes additional information, such as the chemical composition of the product, and which compounds contribute to flavor, is needed. Different instrumental techniques can be used to

Table I: References for the Descriptive Sensory Analysis of SMP and WMP

Descriptor	*Definition*	*Reference*	*Preparation*
Cooked/sulfurous	Sulfurous aromatic associated with heated milk	Fresh UHT milk, cooked milk	Heat pasteurized skim milk to 85°C for 45 min
Cooked/caramelized	Sweet aromatic associated with burnt sugar or butterscotch		Dilute a table spoon of caramel syrup in 400 ml skim milk
Sweet aromatic	Sweet aroma associated with dairy products		Vanilla cake mix or 20 ppm vanillin in milk
Grassy/hay	Aromatics reminiscent of freshly cut grass or hay	Hexanal	1 ppm hexanal in water or fresh cut grass
Feed	Aromatics reminiscent of malted grains	2/3-methylbutanal	1 ppm 2/3 methyl butanal in water or grape nuts cereal in milk
Brothy	Aromatics associated with broth or boiled potatoes	Methional	1 ppm methional in water or boiled potatoes
Milkfat/lactone	Aromatic associated with milkfat	δ-dodecalactone	1 ppm δ-dodecalactone in water or heavy cream
Fatty/fried	Aromatic reminiscent of old fryer oil and fried foods	(E,E)-2,4-decadienal	1 ppm (E,E)-2,4-decadienal in skim milk
Painty	Aromatic reminiscent of solvents and paint		Paint and turpentine
Fishy/doughy	Aromatic reminiscent of fresh fish	(Z)-4-heptanal	1 ppm (Z)-4-heptanal or canned biscuit dough
Cardboard/wet paper	Aroma associated with cardboard	Cardboard paper	Cardboard in water
Grape/tortilla	Aromatic reminiscent of stale tortillas and grape flavoring	*o*-aminoacetophenone	1 ppm *o*-aminoacetophenone in water
Sweet taste	Fundamental basic taste associated with sugars	Sucrose	5% sucrose in water
Salty	Basic taste associated with salts	NaCl	2% NaCl in water
Astringency	Drying tongue sensation	Alum	Alum, 1% in water

Adapted from references *9, 16, 53, 84.*

extract and analyze aroma-active compounds in foods. Because the aroma compounds found in foods are usually present at very low concentrations, isolation and concentration procedures are often needed in order to obtain the chemicals of interest in concentrations that can be detected (*23*). There are several extraction and concentration techniques that are commonly employed to extract the volatile flavor-contributing compounds in foods including direct solvent extraction/high vacuum distillation (DSE/HVD) and solvent assisted flavor evaporation (DSE/SAFE), solid phase microextraction (SPME), and dynamic headspace analysis/gas chromatography (DHA/GC). DSE/HVD or DSE/SAFE, DHA, and SPME are the most widely applied techniques used to isolate milk flavor volatiles (*24*). The extracted volatiles are then identified using gas chromatography/olfactometry (GC/O), and gas chromatography/mass spectrometry (GC/MS). Traditional solvent assisted extraction techniques are often utilized, and these are most useful in extracting the volatile and semi-volatile analytes. Rapid techniques, including SMPE and DHA, are excellent at extracting the very volatile compounds, and it is advised to use a combination of these two techniques to get a good recovery of volatile compounds.

One rapid technique that is often used to isolate and identify compounds is dynamic headspace/gas chromatography (DHA/GC). DHA purges the sample with an inert gas to facilitate the release of volatiles from the food product into the headspace and concentrates the volatiles on to a selective trap. Following concentration, the concentrated volatiles desorbed from the trap onto a GC using a heated transfer line. DHA is a simple and reliable technique that can be used to concentrate and identify trace analytes (*23, 25*). Several studies have effectively utilized this technique with dairy products (*23, 26-29*).

SPME is a rapid technique used in the extraction of volatile compounds (30). A fiber coated with a selective material is inserted into the headspace of a sample for a given time, and the volatile compounds are concentrated onto it. Following exposure, the fiber is injected on to a GC for separation and identification. SPME is inexpensive, solvent-free, and reliable (*31*). SPME has been used in the analysis of dairy products, including cheese and fluid milk (*24, 25, 31*) and has been found to be a useful method in detecting volatile lipid oxidation products including alcohols, ketones, sulfur compounds, fatty acids, and aldehydes (*25*). These volatile extraction techniques are reviewed in more detail elsewhere (*4, 32, 33*).

After volatiles are extracted and concentrated using these methods, gas chromatography/olfactometry (GC/O) is used to separate and tentatively identify the volatiles. A sniffing port is placed on the end of the GC to allow olfactory detection of aroma-active compounds as they are separated on the GC (*34, 35*). GC/O can be used in conjunction with most concentration and extraction techniques including DSE/SAFE, SPME, and DHA. GC/O allows not only chemical information of a compound or sample to be gained, but also gives sensory information and clues into the role that the compound(s) plays in the flavor profile of the product. Though GC/O provides tentative identification of aroma active compounds, GC/MS must be used in conjunction with GC/O to provide a positive identification. Positive identification is confirmed by comparing retention indices and mass spectra with authentic standards.

In this study, volatile compounds from SMP and WMP extracts were separated using a high vacuum distillation technique detailed by Karagul-Yüceer et al. (*21*). The apparatus used was similar to that described by Sen et al. (*36*). The distillation process began by placing the extract into a 1-L round bottom flask and immersing it into a Dewar vessel containing liquid nitrogen until shell frozen. The frozen flask was then immediately connected to a distillation unit equipped with a rough pump/diffusion pump as the vacuum source (about 10^{-4} Torr), a receiving tube, and a waste tube. The receiving tube and waste tube were held in separate Dewar vessels containing liquid nitrogen until distillation was completed (4 h). For the first two h, the sample flask was held at room temperature. During the second two h, the sample was kept thermostated in a water bath at 50 °C. After distillation, the distillate was concentrated to 20 mL under a stream of nitrogen gas. The concentrated distillate was then washed twice with three mL sodium bicarbonate (0.5 M) and vigorously shaken. It was then washed three times with two mL saturated sodium chloride solution. The upper layer (ether) containing the neutral/basic fraction was collected in a glass tube using a pipette. The distilled extracts were then dried over anhydrous sodium sulfate and concentrated to 0.5 mL under a stream of nitrogen gas. Acidic volatiles were recovered by acidifying the bottom layer (aqueous phase) with about five mL of 6.2 M hydrochloric acid to pH 2-2.5 and extracting the sample three times with five mL ethyl ether. The extracted acidic volatiles were then dried over anhydrous sodium sulfate before concentration to 0.5 mL under nitrogen.

This research utilized GC/O to identify and characterize aroma-active compounds in the SMP and WMP extracts. An HP5890 series II gas chromatograph (Hewlett-Packard Co., Palo Alto, CA) equipped with a flame ionization detector (FID), a sniffing port, and a splitless injector was utilized in GC/O. Both the neutral/basic and acidic fractions were analyzed from each duplicate extraction. Two μL were injected onto a polar capillary column (DB-WAX, 30 m length x 0.25 mm i.d. x 0.25 μm film thickness d_f; J. & W. Scientific, Folsom, CA) and a nonpolar column (DB-5MS, 30 m length x 0.25 mm i.d. x 0.25 μm d_f; J & W Scientific, Folsom, CA). Column effluent was split 1:1 between the FID and sniffing port using deactivated fused silica capillaries (one m length x 0.25 mm i.d.). The GC oven temperature was programmed from 40 °C to 200 °C at a rate of 10 °C/min with an initial hold for three min and a final hold of 20 min. The FID and sniffing port were maintained at a temperature of 250 °C. The sniffing port was supplied with humidified air at 30 mL/min. Post peak intensity was used to characterize the odorants in the extracts (*35*). Four experienced panelists each with more than 150 h of experience sniffed the neutral/basic and acidic fractions of cheese extracts twice on the two different columns. Sniffers described the odor and scored the intensity of odorants in the extracts using a 10-point numerical intensity scale (*35*).

For positive identifications, retention indices (RI), mass spectra, and odor properties of unknowns were compared with those of authentic standard compounds analyzed under identical conditions. Tentative identifications were based on comparing mass spectra of unknown compounds with authentic standards or on matching the RI values and odor properties of unknowns against those of authentic standards. For the calculation of retention indices, an n-alkane series was used (*38*).

Various statistical analyses, both univariate and multivariate, were used to analyze the data. Sensory data and instrumental quantification results were analyzed using the SAS statistical software (version 8.2, SAS Institute, Cary, NC). Instrumental data were treated as a completely randomized design with repeated measures. Sensory data were collected in a randomized balanced block with repeated measures (with panelist as the block). Analysis of variance with means separation (least square means) was conducted to identify differences. Principal component analysis (PROC PRINCOMP) was also conducted on the sensory analysis data of different milk powders during aging.

Results

Sensory profiles of fresh and stored milk powders are found in Tables II and III. The only basic tastes that were present were sweet and salty flavors as well as astringency (Tables II and III). In general, astringency increased with storage time (Tables II and III). As the powders aged, the intensity of the off-flavors increased. In WMP, both cooked/sulfur and cooked/caramelized decreased with storage while fatty/fryer oil, grassy, and painty flavors increased. Similarly, in SMP, the cooked and sweet aromatic flavors decreased with storage time, but fatty/fryer oil and grape/tortilla flavors increased. These non-dairy flavors are likely related to lipid oxidation and protein degradation that occurs during storage. As such, they should be considered undesirable or off-flavors.

This study examined flavor during storage of two different SMP and WMP in order to focus on volatile changes of these products and lay a structure for studying flavor stability. Additional studies are currently being conducted with larger samplings of SMP and WMP. Figure 1 demonstrates the large amount of flavor variability that may exist from different fresh SMP. Clearly the freshness of the powder (i.e. early shelf life status) does not guarantee a fresh fluid milk-like flavor profile. There are wide differences in the flavor of these samples, and as these products age, the flavor variability will only increase.

Sixty-five different aroma active compounds were identified in fresh and stored WMP. Twenty two of these compounds were positively identified, thirty-

Table II: Sensory Analysis of Fluid Whole Milk and Whole Milk Powder

Attribute	*Fluid whole*[A]	*Fresh WMP*	*12 month WMP*	*24 month WMP*
Cooked/sulfurous	2.00ab	1.92ab	2.00ab	1.33b
Cooked/caramelized	1.50ab	1.67ab	1.42bc	0.75c
Sweet aromatic	2.03a	2.75a	2.02a	1.08b
Grassy/hay	ND	ND	1.92a	1.33a
Milkfat/lactone	3.00a	2.43a	2.33a	0.83b
Fatty/fryer oil	ND	ND	1.33b	2.42a
Painty	ND	ND	ND	0.72ab
Cardboard	ND	ND	ND	ND
Fishy/doughy	ND	ND	ND	ND
Grape/tortilla	ND	ND	ND	ND
Sweet taste	2.06bc	3.33a	1.83c	2.58b
Salty	ND	0.25a	0.25a	ND
Astringency	1.64a	ND	1.77a	2.05a

[A] Fluid whole pasteurized milk. NOTE: Different letters within rows indicate significant differences ($P<0.05$).

Table III: Sensory Analysis of Fluid Skim Milk and Skim Milk Powder

Attribute	*Fluid Skim*[A]	*Fresh SMP*	*12 month SMP*	*24 month SMP*	*36 month SMP*
Cooked/sulfurous	2.50b	3.25a	2.42b	2.17b	2.25b
Cooked/caramelized	ND	ND	ND	ND	ND
Sweet aromatic	2.00a	2.67a	2.05ab	1.83b	1.45b
Grassy/hay	0.25a	ND	ND	ND	ND
Milkfat/lactone	0.42a	ND	ND	ND	ND
Fatty/fryer oil	ND	ND	0.67ab	1.22a	0.88a
Painty	ND	ND	ND	ND	ND
Cardboard	ND	ND	0.30a	ND	ND
Fishy/doughy	ND	ND	ND	ND	ND
Grape/tortilla	ND	ND	ND	ND	0.70a
Sweet taste	2.95a	2.63a	1.83b	1.92b	2.58a
Salty	0.33a	ND	ND	ND	ND
Astringency	1.67b	1.00c	1.58b	2.33a	1.33bc

[A] Fluid pasteurized skim milk. NOTE: Different letters within rows indicate significant differences ($P<0.05$).

two were tentatively identified and nine of these compounds are unknown (Table IV). To our knowledge, there has been no other study that has examined the flavor of WMP using descriptive sensory analysis in conjunction with GC/O and GC/MS. These techniques provide an excellent way to determine which compounds contribute to aroma in WMP. A few of these compounds have been previously identified (*39, 40*) in studies that utilized instrumental techniques only.

In WMP, there are many different classes of compounds, but many compounds were formed via lipid oxidation (aldehydes, ketones, fatty acids, and esters). In general, the intensity and number of aldehydes (both saturated and unsaturated) increased with storage time (Table IV). Most of these compounds have fatty aromas, and the increase in intensity of these compounds coincides with the increase in fatty/fryer oil flavor in the stored WMP. Though fatty acids and esters may be formed from fat degradation reactions, the intensities of these aroma-active compounds remained fairly constant during storage, suggesting that they may be formed during the WMP production process. There were also several lactone compounds found in WMP. Lactones are formed from milkfat during heating (*41*). Lactones help to give milkfat its characteristic sweet, coconut aroma.

Many Strecker aldehydes and other protein degradation products were also identified in both fresh and stored WMP. Thermally generated compounds such as pyrrolines and thiazolines were also identified. These have popcorn-like aromas, and have been found to contribute to the flavor of fresh milk powders (*1, 21, 42*). There were also sulfur compounds identified in WMP. Methional is important to dairy flavor, and has been previously identified as a key odorant in stored milk powder (*1*). Though cooked/sulfur flavor is important to the characteristic flavor of fluid WMP, with the exception of methional, the sulfur compounds identified in WMP (2-methyl thiophene (plastic/rubber), 2-methyl-3-furanthiol (brothy/vitamin),and benzothiazole (rubbery)) do not seem to contribute as much to the overall flavor of stored WMP as aldehydes. The overall number of compounds that were detected in fresh WMP was 35, while stored WMP contained over 50 different aroma active compounds. The increase in the number of compounds during storage is attributed to both lipid oxidation and the formation of secondary oxidation products as well as protein degradation. The moisture content of WMP and SMP increased during storage. In WMP and SMP, the initial moisture content was 2.1% and 4.0% respectively, and after storage, the moisture content had significantly increased to 3.9% and 4.5% respectively. Increasing water activity has been shown to increase reaction rates and thereby increase the formation of off-flavors (*5, 43*).

There were 68 aroma-active compounds identified in SMP in this study (Table V). Thirty-one compounds were positively identified, 29 were tentatively identified, and 8 are unknown. Most of these compounds have been previously identified. Karagul-Yuceer et al. (*1, 21*) used similar methods and identified many of the same compounds that were identified in this study. Shiratsuchi et

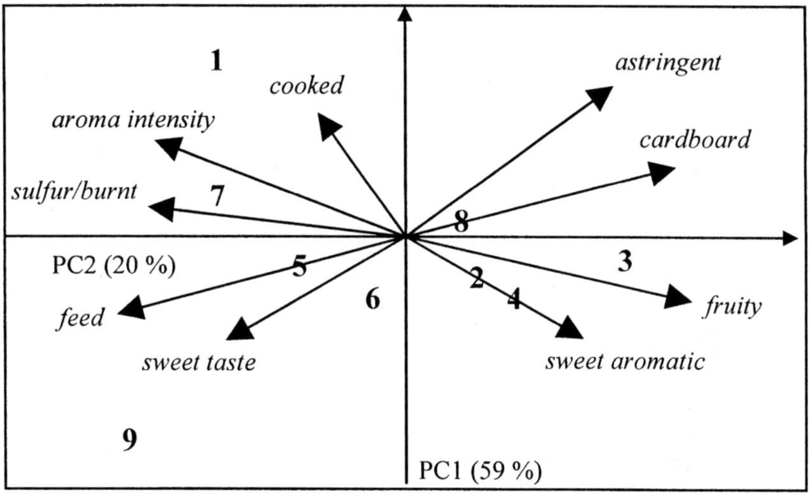

Figure 1. Principal component biplot of the descriptive sensory flavor profiles of rehydrated fresh SMP. [Numbers represent different fresh reconstituted SMP. Samples 1 and 7 are medium heat SMP and samples2-6 and 8-9 are low heat SMP.]

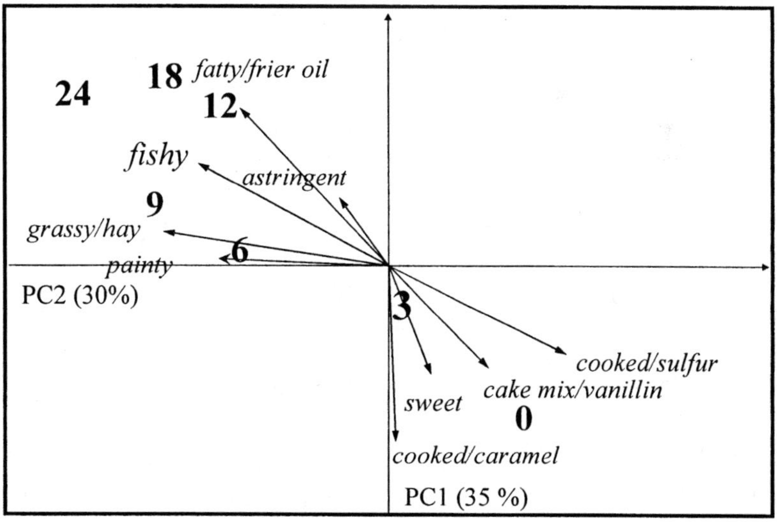

Figure 2. Principle component biplot of descriptive sensory flavor profiles of reconstituted WMP at ambient storage time. [Numbers represent months in ambient storage.]

Table IV. Summary of Odor-active Compounds Identified in Fresh and Stored WMP

compound	*fraction*[a]	*odor*[b]	*RI*[c]		*post peak intensity at the sniffer port (n=4)*				
			DB5	*WAX*	*fresh WMP*	*6 mo WMP*	*12 mo WMP*	*18 mo WMP*	*24 mo WMP*
acetic acid [d]	Ac	vinegar	600	1393	4.0	1.8	1.7	1.0	ND
2-3 butanedione (diacetyl) [d]	NB	buttery	670	956	2.0	1.5	1.5	1.3	1.6
2/3-methyl butanal[d]	NB	malty/chocolate	686	925	ND [f]	ND	ND	ND	1.6
ethyl butanoate[d]	NB	fruity/solventy	730	1000	ND	ND	2.5	1.0	ND
2-methyl thiophene[e]	NB	rubbery/plastic bottle	800	956	3.5	ND	ND	ND	1.1
Hexanal[d,f,g]	NB	green grass	810	1048	ND	3.5	2.6	3.0	3.2
unknown	NB	skunk	827	1067	2.5	2.6	1.7	2.7	1.5
butanoic acid [d]	Ac	cheesy/rancid	830	1610	5.5	4.5	3.7	3.2	1.5
unknown	NB	ammonia	835	1054	ND	3.5	2.3	2.5	2.3
unknown	NB	fruity/solvent	857	1037	ND	ND	ND	1.8	1.7
2-methyl-3-furanthiol [e]	NB	brothy/metallic	875		ND	ND	ND	2.3	1.0
propionic acid [d]	Ac	Swiss cheese	883	1495	2.5	ND	ND	ND	ND
(*Z*)-4-heptenal [e]	NB	fatty/fishy	903	1220	ND	3.5	3.6	3.2	3.4
2/3-methylbutanaoic acid [d]	Ac	sweaty/dried apricots	925	1527	3.5	ND	ND	3.0	ND
pentanoic acid [d]	Ac	sweaty	930	1713	2.5	2.0	2.5	ND	ND
methional [e]	NB	potato	911	1429	1.5	3.8	3.3	3.1	3.0
2-acetyl-1-pyrroline [e]	NB	popcorn	925	1317	6.3	4.8	3.8	3.6	3.3
1-octen-3-one [e]	NB	mushroom	983	1270	2.5	2.0	3.3	3.6	3.5
(*Z*)-1,5-octadien-3-one [e]	NB	geranium	985	1280	ND	ND	2.5	2.3	2.5
(*E,E*)-2,4-heptadienal [e]	NB	fatty	1001	1345	ND	ND	1.8	2.1	2.4
octanal [d]	NB	citrus/green	1008	1273	ND	2.6	2.9	3.3	3.0
hexanoic acid [d]	Ac	sweaty	1019	1861	7.0	2.5	1.9	3.4	1.7
2-acetyl thiazole [e]	NB	popcorn	1043		2.8	ND	ND	ND	2.0
phenylacetaldehyde [d]	NB	rosy/honey	1044	1619	1.0	ND	1.1	2.9	1.5
unknown	NB	coconut	1060		ND	ND	ND	1.7	2.0

(*E*)-2-octenal [d]	NB	citrus/fatty	1066	1345	ND	ND	2.2	1.0	2.3
p-cresol [e]	NB	phenolic	1088	2123	ND	ND	1.4	ND	2.3
2-methoxyphenol (guiacol) [e]	NB	smoky	1092	1480	2.5	1.8	2.3	3.2	2.6
nonanal [d]	NB	fatty/citrus	1100	1385	1.5	2.3	3.0	2.3	3.5
2-acetyl-2-thiazoline [e]	NB	popcorn	1111	1793	2.5	ND	3.1	1.9	ND
maltol [e]	Ac	caramelized	1111	1459	3.8	ND	ND	ND	ND
unknown	NB	fatty/hay		1846	3.0	2.0	1.6	3.2	ND
3-hydroxy-4,5-dimethyl-2(5*H*)-furanone (sotolon) [e]	Ac	maple/spicy	1127	2210	3.6	2.3	2.8	3.1	3.2
unknown	NB	nutty/licorice	1126		2.4	ND	2.5	ND	ND
2-ethyl-4-hydroxy-5-methyl-3(2*H*)-furanone (homofuraneol) [e]	Ac	burnt sugar	1142	2058	3.0	1.8	2.8	2.8	3.5
2-phenethanol [e]	NB	rosy	1150	1427	ND	2.5	3.7	3.6	3.4
(*E*)-2-nonenal [d]	NB	cucumber/old books	1164	1568	2.5	2.6	3.0	4.3	3.4
unknown	NB	earthy/nutty	1174		ND	ND	3.6	4.0	ND
heptanoic acid [d]	AC	sweaty		1904	ND	ND	2.4	1.0	ND
o-cresol [e]	NB	plastic/phenolic	1188		ND	ND	2.5	2.9	2.5
butyl hexanoate [e]	NB	floral/citrus	1203	1536	ND	ND	2.3	1.3	1.5
decanal [d]	NB	fatty	1223	1483	ND	3.0	3.4	2.8	2.5
(*E,E*)-2,4-nonadienal [e]	NB	fatty/hay	1224	1660	1.3	2.2	3.7	4.2	3.5
benzothiazole [e]	NB	burnt rubber	1250	1554	ND	2.0	2.0	2.0	ND
phenyl ethyl acetate [e]	NB	rosy	1260	1820	ND	ND	2.6	2.8	ND
phenyl acetic acid [e]	Ac	honey/floral	1265	2569	ND	ND	ND	3.5	1.5
unknown	NB	cilantro	1268		2.3	1.3	2.7	2.9	3.4
(*E*)-2-decenal [d]	NB	fatty	1270	1586	3.0	ND	ND	ND	ND
δ-octalactone [e]	NB	coconut	1287	1924	5.0	2.8	2.2	2.9	2.6

Continued on next page

Table IV. *Continued.*

compound	*fraction*[a]	*odor*[b]	*RI*[c]		*post peak intensity at the sniffer port (n=4)*				
			DB5	*WAX*	*fresh WMP*	*6 mo WMP*	*12 mo WMP*	*18 mo WMP*	*24 mo WMP*
undecanal [e]	NB	medicinal/fatty	1305		ND	ND	2.5	ND	ND
o-aminoacetophenone [e]	NB	grape/stale	1315	2223	6.0	2.3	3.8	4.2	4.3
(*E,E*)-2,4-decadienal [e]	NB	fatty/oxidized	1325	1726	ND	2.5	3.7	4.7	3.5
4-methyl octanoic acid [e]	Ac	waxy/soapy	1359	2173	ND	ND	3.3	2.9	1.8
γ-nonalactone [e]	NB	coconut	1360	2036	ND	ND	2.3	3.1	2.8
β-damascenone [e]	NB	peach/oatmeal	1386	1689	ND	ND	3.2	3.0	2.8
octanoic acid [d]	Ac	sweaty/waxy	1398	2092	4.0	1.8	2.5	3.1	1.0
3-methyl indole [e]	NB	fecal/mothball	1399	1885	7.0	3.0	3.2	3.3	2.7
unknown	NB	paper/metallic	1425	1938	ND	ND	3.5	4.2	3.2
3-methoxy-4-hydroxybenzaldehyde (vanillin) [e]	Ac	vanilla	1430	1892	ND	ND	ND	2.5	2.6
δ-decalactone [d]	NB	coconut	1481	2210	3.6	2.0	3.0	4.5	2.8
γ-decalactone [e]	NB	peach	1508	2000	ND	2.3	3.4	2.8	3.7
γ-octalactone [e]	NB	coconut	1550	1881	5.8	ND	3.8	4.6	2.0
6-(*Z*)-dodecen-γ-lactone [e]	NB	soapy/sweet	1603		ND	ND	2.5	2.5	3.6
γ-dodecalactone [d]	NB	sweet	1664	2399	5.5	1.0	ND	0.6	ND
δ-dodecalactone [d]	NB	peach	1705		4.0	ND	2.8	3.2	3.8

[a] Ac- acidic fraction, NB – neutral/basic fraction. [b] Odor description at GC-sniffing port. [c] Retention indices (RI) from GC-O data. [d] Compound positively identified by comparison of RI and MS data and odor characteristics with the authentic standard. [e] Compound tentatively identified by comparison of RI data and odor characteristics with authentic standard. [f] Compound not detected.

Table V. Summary of Odor-active Compounds Identified in Fresh and Stored SMP

compound	*fraction*[a]	*odor*[b]	*RI*[c]		*post peak intensity at the sniffer port (n=4)*						
			DB5	*WAX*	*fresh SMP*	*6mo SMP*	*12mo SMP*	*18mo SMP*	*24mo SMP*	*30mo SMP*	*30mo SM*
acetic acid[d]	Ac	vinegar	600	1393	1.5	2.0	1.0	2.6	1.8	1.1	1.0
2-3 butanedione[d]	NB	buttery	670	956	ND[f]	0.5	1.0	1.3	1.0	ND	ND
2/3-methyl butanal[d]	NB	malty/chocolate	686	925	ND	ND	ND	ND	1.0	ND	ND
ethyl butanoate[d]	NB	fruity/solventy	730	1000	ND	ND	ND	2.0	1.0	1.0	ND
2-methyl thiophene[e]	NB	rubbery/plastic bottle	800	956	ND	2.7	3.3	1.0	1.5	ND	ND
hexanal[d,f,g]	NB	green grass	810	1048	ND	2.5	1.9	3.3	1.7	ND	2.0
unknown	NB	skunk	827	1067	ND	ND	1.0	1.8	ND	1.0	ND
butanoic acid[d]	Ac	cheesy/rancid	830	1610	2.3	5.5	2.9	4.0	2.5	2.3	3.1
unknown	NB	ammonia	835	1054	ND	1.0	2.1	3.1	1.8	1.0	2.8
unknown	NB	fruity/solvent	857	*1037*	ND	ND	ND	1.6	1.0	1.3	2.3
2-methyl-3-furanthiol[e]	NB	brothy/metallic	875		ND	ND	1.0	1.6	1.0	ND	ND
propionic acid[d]	Ac	Swiss cheese	883	*1495*	1.0	ND	ND	ND	ND	ND	ND
(*Z*)-4-heptenal[e]	NB	fatty/fishy	903	1220	ND	2.4	2.4	2.5	2.2	1.0	1.0
2/3-methylbutanaoic acid[d]	Ac	sweaty/dried apricots	925	1527	2.0	2.5	ND	ND	1.0	1.9	2.0
pentanoic acid[d]	Ac	sweaty	930	*1713*	1.0	1.0	1.8	3.3	1.5	ND	ND
methional[e]	NB	potato	911	1429	3.8	3.3	3.7	3.8	2.7	2.2	3.4
2-acetyl-1-pyrroline[e]	NB	popcorn	925	1317	5.7	4.8	4.1	3.8	3.3	2.7	2.8
unknown	NB	burnt/mushroom	962	1266	ND	ND	2.3	1.5	ND	ND	ND
dimethyltrisulfide	NB	cabbage/sulfur	973	1362	ND	ND	ND	1.2	2.5	1.8	2.4
1-octen-3-one[e]	NB	mushroom	983	1270	2.5	2.8	2.3	2.9	2.1	1.5	2.0
(*Z*)-1,5-octadien-3-one[e]	NB	geranium	985	*1280*	ND	1.0	1.8	2.7	1.0	ND	2.1
(*E,E*)-2,4-heptadienal[e]	NB	fatty	1001	1345	ND	2.8	1.6	2.5	1.0	ND	ND
octanal[d]	NB	citrus/green	1008	1273	ND	2.0	2.7	2.8	2.9	ND	2.4
hexanoic acid[d]	Ac	sweaty	1019	1861	4.5	2.4	2.3	4.1	1.6	ND	2.5
2-acetyl thiazole[e]	NB	popcorn	1043		1.0	ND	1.9	2.5	2.3	ND	2.0

Continued on next page.

Table V. *Continued.*

compound	*fraction*[a]	*odor*[b]	*RI*[c]		*post peak intensity at the sniffer port (n=4)*						
			DB5	*WAX*	*fresh SMP*	*6mo SMP*	*12mo SMP*	*18mo SMP*	*24mo SMP*	*30mo SMP*	*30mo SM*
phenylacetaldehyde [d]	NB	rosy/honey	1044	1619	2.0	1.8	1.8	2.2	1.0	1.0	2.8
(*E*)-2-octenal [d]	NB	citrus/fatty	1066	1345	ND	ND	2.5	2.6	1.7	ND	ND
2,5-dimethyl-4-hydroxy-3(*2H*)-furanone (Furaneol) [e]	Ac	burnt sugar	1071	2054	4.5	2.8	2.5	3.1	3.3	2.4	3.0
p-cresol [e]	NB	cowy/badaid	1088	2123	ND	ND	ND	2.8	2.8	1.0	ND
2-methoxyphenol (guiacol) [e]	NB	burnt/smoky	1092	1480	ND	ND	2.2	3.0	2.0	1.3	2.6
nonanal [d]	NB	fatty/floral	1100	1385	1.5	ND	2.5	3.5	3.1	ND	2.6
2-acetyl-2-thiazoline [e]	NB	popcorn	1111	1793	1.0	ND	ND	2.6	1.0	ND	1.8
maltol [e]	Ac	burnt sugar	1111	1459	ND	ND	ND	ND	2.0	2.2	3.4
3-hydroxy-4,5-dimethyl-2(5*H*)-furanone (sotolon) [e]	Ac	curry/maple/spicy	1127	2210	2.5	3.3	3.2	3.6	1.8	2.1	2.8
2-ethyl-4-hydroxy-5-methyl-3(2H)-furanone (homofuraneol) [e]	Ac	burnt sugar	1142	2058	2.8	2.8	ND	ND	ND	ND	ND
2-phenethanol [e]	NB	floral/rosy	1150	1427	ND	2.5	2.5	2.2	2.3	1.8	2.7
(*E,Z*)-2,6-nonadienal [e]	NB	cucumber	1154	1573	1.3	1.6	2.6	3.0	2.9	2.2	2.2
(*E*)-2-nonenal [d]	NB	old books/paper	1164	1568	ND	1.7	2.7	3.4	2.4	2.5	2.3
unknown	NB	earthy/nutty	1174		1.0	ND	2.3	1.7	2.7	ND	ND
butyl hexanoate [d]	NB	rosy	1203	1536	ND	ND	2.2	2.4	2.0	ND	ND
decanal [d]	NB	fatty	1223	1483	ND	ND	ND	2.8	2.3	1.0	1.8
(*E,E*)-2,4-nonadienal [e]	NB	fatty	1224	1660	ND	2.4	2.6	2.3	2.5	ND	ND
benzothiazole [e]	NB	rubbery	1250	1554	ND	ND	ND	ND	1.0	ND	2.3
(Z)-2-decenal [d]	NB	fatty/hay	1256		ND	3.0	2.8	2.8	1.4	ND	1.0
phenyl ethyl acetate [e]	NB	rosy	1260	1820	ND	ND	2.5	2.5	1.8	1.0	ND
phenyl acetic acid [e]	Ac	fatty/rosy	1265	2569	ND	ND	ND	3.9	3.5	1.0	ND
δ-octalactone [e]	NB	coconut	1287	1924	4.5	1.0	3.3	3.0	2.1	1.5	2.2
unknown	Ac	sweaty/urine	1150	1590	ND	1.9	ND	ND	ND	ND	ND

undecanal[e]	NB	floral/tobacco/ spicy	1305		1.0	1.0	1.4	2.7	2.7	ND	ND
(*E*,*Z*)-2,4-decadienal	NB	fatty	1316	1621	ND	2.5	2.8	1.6	2.5	2.3	2.2
o-aminoacetophenone[e]	NB	grape/stale	1315	2223	4.5	1.5	3.0	3.9	3.7	2.5	2.4
(*E*,*E*)-2,4-decadienal[e]	NB	fatty/oxidized	1325	1726	ND	1.5	3.4	3.4	2.3	ND	2.2
4-methyl octanoic acid[e]	Ac	waxy/soapy	1359	2173	ND	ND	3.2	3.9	2.2	1.8	1.0
γ-nonalactone[e]	NB	coconut/ cilantro	1360	2036	ND	1.0	2.6	3.4	1.7	1.0	1.0
β-damascenone[e]	NB	peach/oatmeal	1386	1689	ND	ND	ND	3.2	ND	ND	ND
octanoic acid[d]	Ac	sweaty/waxy	1398	2092	2.3	2.8	3.3	3.4	1.0	1.0	1.8
3-methyl indole[e]	NB	fecal/mothball	1399	1885	1.0	ND	1.5	2.3	1.7	ND	1.0
dodecanal	NB	sweet/citrus	1400	1762	ND	ND	2.4	2.9	2.0	1.8	ND
unknown	NB	paper/metallic	1425	1938	2.0	2.4	3.1	2.6	3.5	3.0	1.5
3-methoxy-4-hydroxybenzaldehyde (vanillin)[e]	Ac	vanilla	1430	1892	ND	2.5	ND	2.9	1.0	1.3	2.0
unknown	Ac	honey	1470	<2400	ND	3.8	ND	2.2	1.5	ND	1.8
δ-decalactone[d]	NB	coconut	1481	2210	ND	2.5	2.0	3.1	1.8	1.3	1.8
γ-decalactone[e]	NB	coconut	1508	2000	ND	2.5	3.0	3.5	2.2	1.4	1.5
γ-octalactone[e]	NB	coconut	1550	1881	3.5	ND	ND	ND	1.0	1.0	2.0
6-(*Z*)-dodecen-γ-lactone[e]	NB	soapy/sweet	1603		2.5	ND	2.9	2.9	1.4	ND	1.6
γ-dodecalactone[d]	NB	sweet	1664	2399	3.8	2.0	ND	ND	1.2	1.8	1.8
δ-dodecalactone[d]	NB	peach	1705		ND	ND	2.5	3.5	3.3	ND	1.3
nonanoic acid[d]	Ac	sweaty		2053	ND	1.0	1.9	2.6	1.0	1.9	2.3

[a] Ac- acidic fraction, NB – neutral/basic fraction. [b] Odor description at GC-sniffing port. [c] Retention indices (RI) from GC-O data. [d] Compound positively identified by comparison of RI and MS data and odor characteristics with the authentic standard. [e] Compound tentatively identified by comparison of RI data and odor characteristics with authentic standard. [f] Compound not detected.

al. (*44*) identified some volatile compounds in SMP by GC/MS only. As with WMP, there were many lipid oxidation compound identified. However, these compounds were present (in general) at lower intensities in SMP than in WMP. There was a large increase in the number of compounds in SMP during storage time. There were 27 compounds identified in fresh SMP and over 60 identified in stored SMP. Interestingly, the 24 mo samples had the highest number of aroma-active compounds and also the highest intensity of fatty/fryer oil flavor. As the SMP continued to age beyond 24 mo, there was a decrease in the number of aroma-active compounds and the overall flavor intensity was lower. This could be caused by the continued breakdown of lipids and proteins into larger and less aroma-active compounds that do not contribute flavor as greatly as the lower molecular weight compounds with low threshold values. Volatile components might also be scalped by the packaging materials.

In general the same compounds were identified in both SMP and WMP. It is expected that the WMP would have a greater number of lipid oxidation products due to its higher lipid content, however there were very little differences in the identity of the aroma-active compounds in SMP and WMP. At first glance it might seem strange that there were a similar number of aroma-active compounds identified in SMP and WMP, the intensity of the compounds identified in WMP was greater than the compounds identified in SMP. This may mean that there are more compounds contributing to SMP flavor, but they are present at lower concentrations, perhaps even below aroma threshold in the powder matrix. There are many different points during the manufacture and storage of SMP and WMP in which flavors may be formed. It is important to look at the starting composition of milk as well as the processing parameters to fully understand milk powder flavor.

Discussion: Flavor Formation

Fluid milk composition and flavor

When trying to understand the flavor of milk powders, it is important to first look at the starting material, i.e. fluid milk. Milk is a diverse mixture containing lipids, protein, and carbohydrates in combination with minerals, gasses, and hormones (*45*). The composition of milk is approximately 87% water, 5% lactose, 4% fat, 3% protein, and 1% ash (*46*). The degradation of protein and lipid are the primary sources of flavor formation in dairy products (*47*).

Caseins make up >80% of the total protein found in milk at a concentration of (24-28g/L) (*48*). Whey proteins make up ~20% of the total protein in milk. β-Lactoglobulin (β-lg) is the major whey protein in milk. It contains two disulfide bridges and a free thiol group (*46*). During heating, the protein begins to unfold, exposing cysteine and cystine groups and hydrophobic surfaces (*49*). Sulfur flavor is very important to pasteurized fluid milk (*50*), and these sulfur containing amino acids contribute to the characteristic flavor of pasteurized fluid milk. Not only does heating influence the flavor of fluid milk, heating (and subsequent protein denaturation) could contribute to the flavor of SMP and WMP during both the pasteurization of milk and during spray drying.

Most of the lipid in bovine milk fat is triacylglycerols (96-98%) (*46*). There are 400 different fatty acids found in milk fat, but only 14 are present in concentrations greater than 1% (*51*). There are 2744 isomers of these 14 fatty acids, and they can have different flavor profiles (*48*). Milk also contains 360mg/kg of free fatty acids, especially low molecular weight (*41*). These short chain free fatty acids have a relatively low aroma threshold and contribute to flavor (*52, 53*). Bovine milk is unique in the fatty acid profile, and is the primary milk source utilized in the production of milk powders. In minor species milk (ovine and caprine), there is a higher concentration of branched chain fatty acids, which contribute to the animal/waxy flavors found in products made with these milks (*54*).

The vitamin and mineral content can also potentially be a source of flavor development in milk and milk products. Milk contains about 1.75 mg/L of riboflavin (*55*). Dimick (*55*) and Aurand et al., (*56*) showed that riboflavin acts as a catalyst in the photo-oxidation of milk fat. Riboflavin is yellow, and there is a strong correlation between the disappearance of the yellow color (and presumably the riboflavin) and the appearance of a brothy flavor (*57*). Riboflavin is thought to be a major contributor to the light-oxidized flavor of milk. Light-oxidized flavor is very common in commercially packaged fluid milk (*58*). The copper content of milk is also important in oxidation reactions because copper acts as a catalyst for lipid oxidation and the formation of off-flavors (*59*). Additionally, natural emulsifiers in milk, such as lecithin, cephalin, and sphingomyelin, are associated with proteins and are easily oxidized in fluid milk (*59*).

Based on the composition of milk, there are obviously several sources of flavor development in milk and subsequent products. There is a sweetness associated with milk from the lactose and this is balanced with a slight salty flavor derived from the salts present in milk (*50*). Sodium and potassium chloride are common salts found in milk that contribute to salty taste (*50*). Other salts, including phosphate and citrate are important, but do not directly contribute to salty flavor (*60*). Milk has a very delicate flavor balance, and if one compound is unbalanced, off-flavors may become apparent (*41*). Often the compounds that are contributing to flavors are present at or below instrumental threshold and are difficult to determine using instrumental techniques (*41*).

There have been many studies examining the flavor of fluid milk, but in general raw, fresh milk has a bland but characteristic flavor (*8*). There are at least 400 volatile compounds that have been detected in raw milk (*61*), but most of them are present in very low concentrations. Charm analysis (a GC/O technique) was conducted on raw milk, and ethylbutanoate (fruity), ethylhexanoate (fruity) and dimethylsulphone (sulfur) were found to be the most aroma active compounds in raw milk (*62*). Calvo and de la Hoz (*63*) stated that there are four things that lead to the characteristic flavor of milk: compounds from the cows' metabolism, compounds from chemical reactions, enzymes or microflora prior to processing, pasteurization, and compounds formed during storage. In raw milk, most flavors are formed during animal metabolism, though feed is also thought to play a key role. The direct transfer of flavor volatiles from the feed to the milk may also take place (*8*).

(*Z*)-4 Heptenal (doughy/fatty), 1-octen-3-one (metallic/mushroom), and hexanal (cut grass) are commonly found in fresh milk (*50*). Dimethyl disulfide (onion/sulfur) is also an important aroma compound in fresh milk (*50*). Other compounds in milk, including carbonyls, alkanols, free fatty acids, and other sulfur compounds are likely produced by the cows' metabolism (*61*). The proper storage of raw milk is important. Raw milk is transported and may be stored in bulk at refrigeration temperatures. The storage time of raw milk has a concrete effect on the quality and flavor of the dried milk powder. Free fatty acid content was greater in powder made from stored milk than powder made from a fresh raw control (*64*).

The cows' feed can also influence the flavor of the milk and milk products. Pasture fed cows have fewer lactones in their milk than silage fed cows (*65*). Milk from pasture fed and total mix ration fed cows have similar flavor compounds present, but the general ratio of these compounds is different (*66*). Badings (*61*) and Bendall (*66*) found that a combination of pasture feeding led to increased concentrations of indole, skatole, sulfides, mercaptans, nitriles and thiocyanates which lead to off-flavors in fluid milk. Pasture fed cows have more protein in their diet and this can influence the flavor of the milk and resulting milk products (*66*). Further discussion of protein degradation and the subsequent formation of flavor compounds can be found later in this paper.

The pasteurization of milk is important not only from a safety perspective, but also because many flavor compounds are formed during heat treatment. The thermal degradation of proteins and lipids can form sulfur compounds, methyl ketones, and lactones (*41, 67*). Hydrogen sulfide is the main sulfur compound in heated milk and is formed from degradation of sulfur containing amino acids, especially in β-lg (*63*). These sulfur containing amino acids form aroma active compounds and contribute to the flavor of SMP and WMP during the pasteurization of fluid milk and subsequent spray drying.

Fresh milk doesn't have any lactone aroma, but after pasteurization, lactones are formed (*65*). Low levels of lactones may be formed from microbiological pathways in raw milk, but higher levels of lactones can be formed from thermal generation. This is why milk processed using UHT conditions (4.6 sec at 142 °C) had a higher intensity of aroma compounds, including lactones, than milk processed using lower temperatures (73 °C for 12 sec). This is caused by Maillard reactions and the thermal oxidation of lipids

(*41*). Heating may also form flavors by increasing the reactivity of unsaturated fatty acids. Though most unsaturated fatty acids are fairly stable, heating the milk during pasteurization may rearrange the double bonds and increase the reactivity of these fatty acids (*56*).

Proper storage of freshly pasteurized milk is very important in controlling flavor and minimizing both variability and off-flavor formation. The storage conditions of milk are very important to the flavor, and consequently the products made from milk. Milk that is processed into powder can be held up to 2 d prior to drying (*68*). The storage time of fluid milk have a definite effect on the quality and flavor of the dried milk powder. Fluid milk may become light oxidized and cause flavor changes (*55, 57*). Milk exposed to sunlight for short periods of time (10-15 minutes) had brothy notes (*69*) as well as metallic or tallow-like flavors (*61*).

Lipolysis and proteolysis can occur in the fluid milk prior to processing into powder. Lipolytic rancidity can be initiated by either milk lipase or bacterial lipases (*61*). Psychotrophic bacteria can also produce esters of butyric, isovaleric, and hexanoic acids which can lead to fruity off-flavors (*8*). A fruity off-flavor was noted in SMP (See Figure 1).

SMP and WMP flavor and composition

The processing of SMP and WMP is another potential source of flavor formation. As the water is removed during concentration and spray drying, the composition of whole milk powder in the US is 24.5-27 % protein, 26-28 % fat, 2-4.5 % moisture, 36-38.5 % lactose, and 5.5-6.5 % ash (*2*). SMP, by law, is comprised of 34-37 % protein, 0.6-1.25 % fat, 3-4 % moisture, 49.5-52 % lactose, and 8.2-8.6 % ash (*2*). The composition of these powders is such that there is a high concentration of protein and lactose. These factors, combined with heat during spray drying, allow for the potential of Maillard browning to occur and the resultant flavor formation. Additionally, in WMP, there is a high concentration of lipid and ash. As previously stated, minerals can act as pro-oxidants and increase the formation of volatile lipid oxidation products. More detail will now be given to these reactions.

SMP and WMP are used primarily as ingredients in other food applications, and this means that they are commonly stored. SMP stored under optimal conditions have a shelf life of anywhere between 12-36 months for non-instantized unfortified SMP under optimal storage conditions (*2, 7, 10*). A recent study suggested that SMP packaged in cans might be acceptable for emergency use after as long as 29 y of ambient storage (*70*). The shelf life of WMP is between 6-9 months under optimal conditions (*2*) though this varies widely with storage conditions.

The flavor of both SMP and WMP has been examined extensively. Flavor variability in SMP has been documented by trained panelists. Drake et al. (*9*) developed a defined sensory language for SMP and documented wide flavor variability among fresh (<3 mo) and stored (> 6mo) reconstituted SMP. Driscoll et al. (*71*) evaluated sensory quality of SMP under various storage conditions. Some descriptive sensory analysis was used, but with undefined and ambiguous attributes. They reported an overall decrease in milk powder quality, as reported by 11 trained panelists, during storage as a direct reflection of the development of an oxidized/stale flavor. Animal-like, fatty/stale, and fecal flavors were observed in stored skim milk powders (*1*). In this study, off-flavors began to develop quickly in the milk powders (Figures 2 and 3). A subsequent study currently is underway and is investigating several SMP during storage. Preliminary data confirmed that variability was found in the sensory stability of SMP, and some develop off-flavors as quickly as 6 months.

Protein degradation (both proteolytic and non-enzymatic browning) and lipid oxidation are the two main flavor reactions in dairy products (*47, 72, 73*). Interestingly, both bitterness and astringency have been associated with proteolysis of milk proteins (*74, 75*). Astringency is associated with the breakdown of casein (*74*). In highly heated milk, astringency is linked to whey proteins, calcium phosphate, and casein interactions (*74*). Peptides can also contribute to bitterness (*76*). Peptides consisting of non-polar amino acids tend to be bitter (*77*). In most milk powders, the only basic tastes that are present are sweet and slight salty flavors as well as the feeling factor astringency (*1, 9, 21*). In this study, astringency increased with storage time. During prolonged storage, there is likely to be more proteolysis and therefore it is logical that astringency would increase with storage time.

Proteolysis is especially important in aged products such as cheese, and stored products such as milk powders. During storage of milk powders, there can be an increase in proteolysis, leading to increased formation of off-flavors. The breakdown of proteins by enzymes creates peptides and amino acids. These compounds can be further broken down to form volatile compounds. The enzymatic degradation of proteins allows the resulting peptides and amino acids to react with lactose, forming Maillard reaction products. Milk powders contain high concentrations of lactose, which makes them highly susceptible to the Maillard reaction (*78*). Lactose crystallization due to temperature and humidity changes during storage of milk powders can affect the amount of lactose available for the Maillard reaction (*43*). In fresh powder, the lactose is in a metastable state as a glass, but during storage is transformed into stable crystals (*43*). Lactose crystallization can increase water activity (which in turn increases lactose crystallization) and initiate deteriorative chemical reactions (*5, 43*). In this study, the moisture content increased during storage as did off-flavors in both SMP and WMP.

The Maillard reaction and Strecker degradation of alpha amino acids form flavor compounds with distinct aromas and low flavor thresholds (*79*). Maillard reaction products are common in products that have been dehydrated by heat, such as SMP and WMP (*80*). When lactose and the free amino groups react,

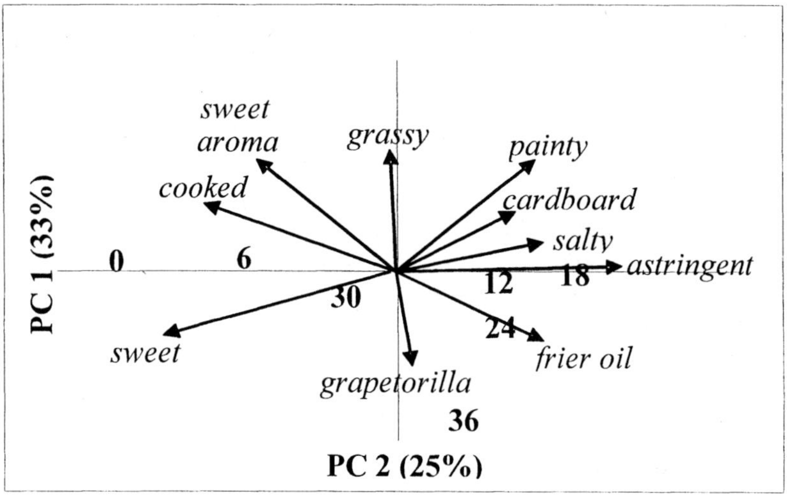

Figure 3. Principle component biplot of descriptive sensory flavor profiles of reconstituted SMP at ambient storage temperatures. [Numbers represent months in ambient storage.]

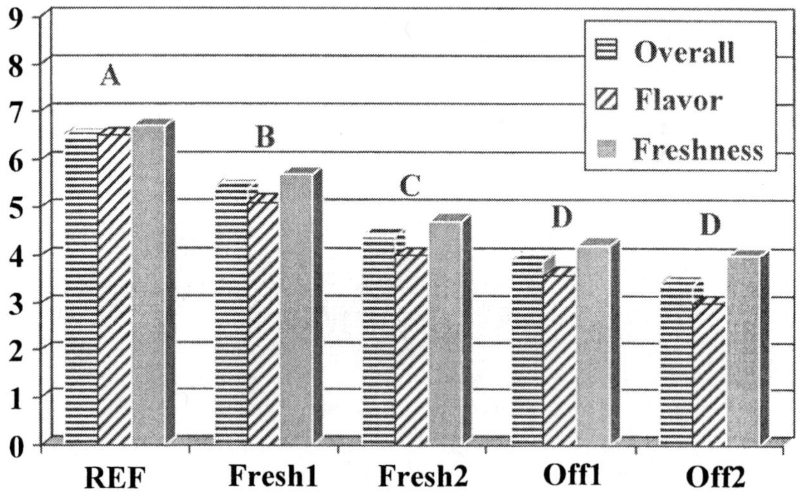

Figure 4. Consumer sensory acceptance of rehydrated SMP. [REF is fluid skim milk, Fresh 1 and 2 are SMP with fresh fluid milk type flavors and Off 1 and 2 are SMP with off-flavors. Descriptive sensory analysis with a trained panel was used to select SMP with fresh- and off-flavors. Different letters indicate significant differences for all attributes (P<0.5) (88).

many flavor compounds are formed with cooked and caramelized flavors (*79*). Maillard reaction products include furans, pyrones, and sulfur compounds (*81*).

Non-specific proteolysis of hydrophobic amino acids such as leucine, phenylalanine, and valine can occur (*82*). The Strecker degradation of aromatic amino acids can form phenylacetaldehyde and phenyl acetic acid which cause rosy flavors in Cheddar cheese (*83*). These compounds are also present in whey proteins (*84*) and were identified in WMP and SMP in this study. The compounds phenylacetaldehyde (honey/rosy), phenyl acetic acid (honey) 2-phenylethanol (rosy) and phenyl ethyl acetate are present in SMP and WMP and all have rosy aromas. The breakdown of aromatic amino acids can also form 2-methoxy phenol, which has a smoky aroma. Tables IV and V summarize compounds that were identified in SMP and WMP this study as well as other studies. The Strecker degradation of other amino acids causes malty/nutty flavor in Cheddar cheese (*85*). These compounds (2-methyl propanal and 2- and 3-methyl butanal) have been identified in SMP (*1, 21*) and WMP (*39*) and may contribute to feed-like flavors or storage-associated off-flavors.

Sulfur compounds can be formed from the degradation of whey proteins (*61*). Methional (potato/brothy) has been identified in both fresh and stored SMP (*1, 21*) and is a Strecker degradation product of methionine. This compound was found in fresh low, medium, and high heat treated SMP. Methional can also cause brothy flavors in Cheddar cheese (*86, 87*). Additionally, methionine can also breakdown into dimethyl disulfide, which is a potent odorant in milk (*50, 87*). Caudle et al. (*88*) found that dimethyl trisulfide and dimethyl disulfide were present in SMP that had been stored and were associated with off flavors by consumers and trained descriptive sensory panelists. Though the sulfur compound 2-methyl-3-furanthiol (brothy/burnt) has not been found in milk powders prior to this study, it has been identified in Cheddar cheese and dried whey proteins (*83, 84*) and is formed from the degradation of sulfur containing amino acids (*89*). 2-methyl thiophene has been previously identified in yogurt and milk (*90*).

Karagul-Yuceer et al. (*1*) identified benzothiazole (rubbery) as an odor-active compound in stored SMP. This is a thermally generated compound formed via degradation of sulfur containing amino acids. o-Aminoacetophenone (grape/tortilla) has been identified as a potent odorant in stored milk powders (*1, 21*). Parks et al. (*91*) identified o-aminoacetophenone and benzothiazole as major contributors to staleness in dried milk and Ramshaw and Dunstone (*92*) determined that o-aminoacetophenone was a key contributor of glue-like staleness in old casein. Grape/tortilla flavor was documented in the stored SMP in this study, and it is likely that *o*-aminoacetophenone is contributing to this flavor.

The thiazolines 2-acetyl-2-thiazoline (popcorn) and 2-acetyl thiazoline (popcorn) were identified in SMP and WMP and are formed from the breakdown of cysteine during heating (*86, 93*). Caseins contribute to most of the free amino acids in milk (*94*), especially proline, histadine, lysine, tyrosine, and aspartamine. Proline is a precursor of 2-aceytl-1-pyrroline (popcorn) (*95*). The reaction between the amino acids proline or ornithine and the reactive sugar-degradation product 2-oxopropanal generates 2-acetyl-1-pyrroline (*95*), which is an important aroma-active compound in fresh SMP and WMP (*1, 21, 42*). 3-Methyl indole (fecal/mothball) was also identified in both SMP and WMP. This compound is formed from the degradation of tryptophan (*66*) and is found in higher intensities in pasture fed cows' milk.

2.5-Dimethyl-4-hydroxy-3-(*2H*) furanone (Furaneol™) (cotton candy) has also been identified in SMP and WMP and is formed during the thermal treatment of sugars. Related compounds 4,5-dimethyl-3-hydroxy-2-(*5H*)-furanone (sotolon) (maple/curry), 2-ethyl-4-hydroxy-3(*2H*)-furanone (homofuraneol) (burnt sugar) and 2-methyl-3-hydroxy-*4H*-pyran-4-one (maltol) (burnt sugar) were also identified in fresh SMP and WMP (*21*). Karagul-Yüceer et al. (*42*) determined that maltol was one of the key contributing volatile compounds to the flavor of fresh SMP. Caudle et al. (*88*) also found maltol in higher relative abundances in fresh flavored SMP compared to off-flavored SMP.

Lipid oxidation reactions are also very important in flavor formation, especially in high fat WMP. Not only does the oxidation of milkfat produce many volatile compounds, the presence of lipids also modifies the environment of Maillard reactions, and can therefore change the resulting products (*80, 96*). One theory is that lipid oxidation and Maillard browning are synergistic in nature, especially in dried dairy products. Intermediate and final products from one of the reactions might increase the rate of the other reaction (*47*).

Lipid oxidation increases the amount of straight chain aldehydes, such as hexanal and this contributes to off-flavors (*47*). The alkanals and alkenals with more than six carbon atoms as well as ketones are the typical volatile lipid oxidation products generated (*44, 86*). These short chain aldehydes have been identified in both SMP (*44*) and WMP (*40*). The oxidation of unsaturated fatty acids forms aldehydes and ketones (*61*). Hammond (*77*) indicated that flavors of oxidized milk arise primarily from small amounts (~3%) of polyunsaturated fatty acids.

Hexanal is used as an indicator of lipid oxidation (*97*). During storage, the concentration of hexanal increased, as did the intensities of off-flavors. Tables II and III show the sensory results from this study, and it is clear that there are increasing off-flavors during storage. Increasing intensities of the descriptors like-casein and like-cardboard were strongly related to increasing concentrations of straight chain aldehydes (*39*). C_5-C_9 aldehydes are key by-products of unsaturated fatty acid oxidation (*98*). The unsaturated aldehydes 2-nonenal and 2,4-heptadienal are secondary products of lipid oxidation (*98*). We identified 2-nonenal by solvent extraction/GC/O, as well as other alkenals and dienals (Tables IV and V) in both fresh and stored SMP and WMP.

Both *p*- and *o*-cresol were identified in WMP and *p*-cresol was identified in SMP. Phenolic compounds are formed both as a result of the degradation of unsaturated fatty acids (*66*) as well as from the degradation of aromatic amino acids (*99*). These compounds smell phenolic/medicinal and are partially responsible for the characteristic flavor of British Farmhouse Cheddar cheese (*100*).

1-Octen-3-one (metallic/mushroom) is formed from the oxidation of arachidonic acid (*86, 87*). (*Z*)-4 Heptenal has a doughy/fishy aroma and was found in stored SMP and WMP (Table IV and V). During storage, the intensity of this compound increased, as did fatty/fryer oil flavor in the powders. The terms oxidized, cardboard, metallic, fatty (oily), painty, and fishy all describe dairy products that have undergone lipid oxidation (*9, 21*). Other alkenals and dienals identified include (*Z*)-4-heptenal (fatty/doughy), (*E,E*)-2,4-heptadienal (fatty), (*E*)-2-octenal (citrus/fatty), (*E*)-2-nonenal (old books), (*E,E*)-2,4-nonadienal (fatty/hay), (E-2-decenal (fatty), (*E,E*)-2,4-decadienal (fatty), (*E,Z*)-2,4-decadienal (fatty), and (*E,Z*)-2,6-nonadienal (cucumber). This compound is an oxidation product of linoleic acid (*86*). 1-octen-3-one and (*E,E*)-2,4-decadienal have been previously identified as primary odorants in milk and milk products (*1, 21, 42, 87*). Other aroma active lipid oxidation product identified in fresh SMP and WMP include hexanal (cut grass), octanal (citrus/fatty), and nonanal (citrus/fatty). These compounds have been previously identified in dairy products (*1, 4, 21, 84*).

In addition to the auto-oxidation of milk fat, milk lipases can cause oxidation as well (*77*). However, these are typically inactivated at pasteurization temperatures and this means that there must be some other means in which the oxidation reaction is initiated, typically by light (*77*). There is also the potential for psychrotrophic microbial degradation of milk fat (*77*). The hydrolysis of triaceylglycerols leads to the formation of free fatty acids. The short chain free fatty acids (C_4-C_{12}) are important flavor contributors of cheese and other dairy products (*4, 87*). Acetic acid (vinegar), butanoic acid (cheesy), pentanoic acid (sweaty), hexanoic acid (sweaty), heptanoic acid (sweaty), octanoic acid (sweaty/waxy), and nonanoic acid (sweaty) were all identified in SMP and WMP. Butanoic and hexanoic acids were noted as the major free fatty acids in skim milk powders (*72*). 2/3-Methyl butanoic acid was identified in SMP (*1, 21*) and has a cheesy/dried apricot aroma. The branched chain fatty acid 4-methyl octanoic acid, which was identified in both SMP and WMP in this study, was not previously identified in milk powders. This compound is not found in high concentrations in cows' milk, but is found in high concentrations in goat and sheep milk (3, 223, 80 ppm respectively) (*101*). Carunchia Whetstine et al. (*102*) found that this acid, in conjunction with 4-ethyl octanoic acid, was the source of waxy flavor in goat cheese. This compound is also found in dried whey proteins (*84*). Several esters were identified in SMP and WMP. These have a fruity aroma and are formed from the esterification of fatty acids (*66*).

Lactones are breakdown products of triglycerides (*103*). Lactones have coconut and peach flavors (*1, 21*) that are characteristic of fresh sweet cream and are also found in young Cheddar cheese (*65, 104*). The mechanism of lactone formation is non-oxidative and the precursors of these compounds are δ-hydroxy fatty acids esterified in milk fat (*65*). Though microorganisms can initiate this pathway and lactone formation, fresh milk doesn't have lactone aroma because lactone concentration is below sensory threshold. However, after pasteurization, lactones are formed above sensory threshold levels, and directly contribute to flavor of freshly pasteurized milk (*65*). 3-Methoxy-4-hydroxybenzaldehyde (vanillin) originates (in the cow rumen) from plant lignin. During pasteurization, this compound is formed (*105*).

It is obvious that storage time of SMP and WMP increases the number of aroma active compounds as well as the intensity of off-flavors. Since the majority of milk powders are used as ingredients, this begs the question can off-flavors from powders carry through into product applications? The quality of SMP has a direct impact on the quality of the finished product (*7*). Hough et al. (*11*) looked at the relationship between consumer acceptability and trained panelists in the evaluation of reconstituted whole milk powder. They found that linear correlations between consumer acceptability and trained sensory panel scores could be used to determine the sensory failure cut-off point and that rejection of whole milk powder was driven by flavor characteristics rather than appearance. Caudle et al. (*88*) found that consumers could detect differences between fresh SMP and off-flavored SMP (Figure 4). They also found that consumers could detect differences in products (vanilla ice cream and hot cocoa) made with fresh and off-flavored SMP (Figures 5 and 6). Products made with off-flavored SMP received lower consumer acceptance scores than products made with fresh-flavored SMP.

There is a seemingly synergistic effect of proteins and fatty acids to increase the rate of flavor formation in dairy products (*47*). It seems that there is no simple mechanism that allows volatiles (and flavors) to form, but rather many working together (*59, 106, 107*). In order to control flavor development, high quality milk and extra care during processing and storage of milk powders needs to be taken.

There are many compounds that contribute to the characteristic flavor of SMP and WMP. During storage, undesirable flavors increase. Aldehydes, ketones, free fatty acids, and alcohols, and lipid oxidation products as well as Maillard reaction products, were associated with undesirable flavors in stored SMP (*1*). Lipid oxidation reaction products, specifically hexanal, are found in higher concentrations in stored powders than in fresh ones (*40*).

2-Acetyl-1-pyrroline (popcorn), Furaneol™ (burnt sugar), butanoic acid (cheesy) and methional are all important to the flavor of fresh SMP and WMP. *o*-Aminoacetophenone (grape/tortilla), 1-octen-3-one (metallic/mushroom), and alkenals and dienals are important flavor contributors to off-flavor in stored milk powders. Because off-flavors from milk powders can carry through into product applications, it is important to understand when and how flavor develops to maximize the potential usage of these functional and important ingredients.

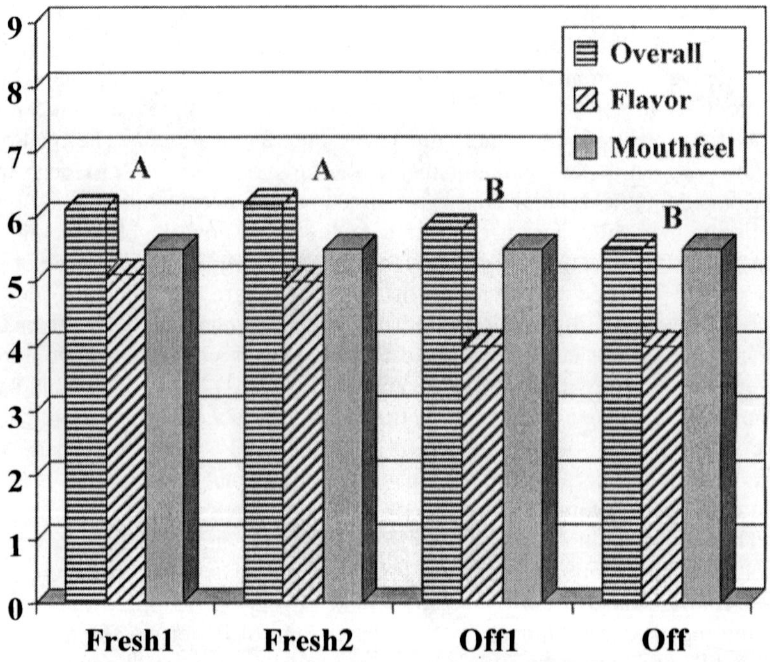

Figure 5. Consumer acceptance scores of vanilla ice creams made with a standard formula. [Ice creams were made with fresh or off-flavored SMP. Fresh 1 and 2 are ice creams made with fresh fluid milk types of flavors and Off 1 and 2 are ice creams made with SMP that exhibited off-flavors. Different letters indicate significant differences from overall acceptance and flavor acceptance (P<0.05). Mouthfeel acceptance was not different (P<0.05) (88).]

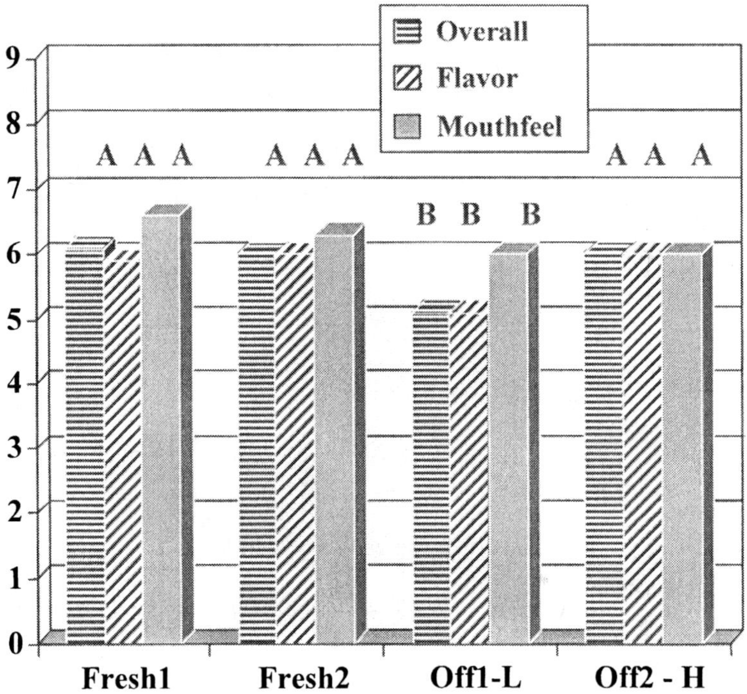

Figure 6. Consumer acceptance scores of reconstituted hot cocoa mix made with a standard formula. [Hot cocoas were made with fresh or off-flavored SMP. Fresh 1 and 2 are hot cocoas made with fresh fluid milk types of flavors and Off 1 and 2 are hot cocoas made with SMP that exhibited off-flavors. Different letters indicate significant differences for overall acceptance, flavor acceptance, and mouthfeel acceptance ($P<0.05$) (88).]

Acknowledgements

Funding provided in part by the California Dairy Research Foundation and Dairy Management, Inc. This is manuscript FSR 06-04 of the Department of Food Science, North Carolina State University. The use of trade names in the publication does not imply endorsement by these organizations or criticisms of ones not mentioned.

References

1. Karagul-Yüceer Y.; Cadwallader K.R.; Drake M.A. *J. Agric. Food Chem.* **2002**, *50,* 305-312.
2. USDEC. Products, Suppliers & Marketing Services for Milk Powders. United States Dairy Export Council. Online. Available: http://www/usdec.org/;products/MilkPowderSpecs/content. Accessed March 25, 2005.
3. Bodyfelt F.W.; Tobias J.; Trout G.M. In *The Sensory Evaluation of Dairy Products;* Van Nostrand Reinhold: London, UK, 1988; pp 67-89.
4. Singh, T.K.; Drake, M.A.; Cadwallader, K.R. *Compr. Rev. Food Sci. Food Safety.* **2003**, *2,* 1-23.
5. Caric, M. *Concentrated and Dried Dairy Products.* VCH Publishers: New York, NY, 1994; pp 57-124.
6. Verdurmen, R.E.M.; de Jong, P. In *Dairy Processing Improving Quality*; Smit G., Ed.; Woodhead Publishing Limited and CRC Press: Boca Raton, FL, 2003; pp 333-365.
7. Varnum A.H.; Sutherland J.P. In *Milk and Milk Products*; Chapman and Hall: New York, NY, 1994; pp 103-149.
8. Boelrijk, A.E.M. In *Dairy Processing Improving Quality*; Smit, G., Ed.; Woodhead Publishing Limited and CRC Press: Boca Raton, FL, 2003; pp 130-154.
9. Drake M.A.; Karagul-Yüceer Y.; Cadwallader K.R.; Civille G.V.; Tong P.S. *J. Sensory Stud.* **2003**, *18,* 199-208.
10. Muir, D.D.; Banks, J.M. In *Dairy Processing Improving Quality;* Smit, G., Ed.; Woodland Publishing and CRC Press LLC: Boca Raton, FL, 2003; pp 201-202.
11. Hough G.; Sanchez R.H.; Garbarini de Pablo G.; Sanchez R.G.; Calderon Villaplana S.; Gimennez A.M.; Gambaro A. *J Dairy Sci.* **2002**, *85,* 2075-2080.
12. USDA. United States Department of Agriculture. Online. Available: http://www.ams.usda.gov/dairy. Accessed March 25, 2005.

13. Delahunty, C.M.; Murray, J.M. In *Proceedings of the 5th Cheese Symposium;* Cogan, T.M., Fox P.F., Ross, R.P. Eds. Dairy Products Research Centre: Fermoy, 1997; pp 90 - 97.
14. Drake, M.A. *J. Dairy Sci.* **2004**, *87,* 777-784.
15. Lawless, H.T.; Heymann, H. *Sensory Evaluation of Food: Practices and Principals;* Chapman and Hall: New York, 1998.
16. Meilgaard, M.C.; Civille, G.V.; Carr, B.T. *Sensory Evaluation Techniques, 3rd Edition;* CRC Press: Boca Raton, FL, 1999.
17. Murray, J.M.; Delahunty, C.M.; Baxter, I. *Food Res. Int.* **2001**, 34, 461-471.
18. Drake, M.A.; Civille, G.V. *Compr. Rev. Food Sci. Food Safety* **2003,** *2,* 33-40.
19. Drake, M.A.; Gerard, P.D.; Wright, S.; Cadwallader, K.R.; Civille, G.V. *J. Sensory Stud.* **2002**, *17,* 215-229.
20. Drake, M.A.; Yates, M.D.; Gerard, P.D.; Delahunty, C.M.,; Sheehan, E.M.; Turnbull, R.P.; Dodds, T.M. *Int. Dairy J.* **2005**, *15,* 473-483.
21. Karagul-Yüceer Y.; Drake M.A.; Cadwallader K.R. *J. Agric Food Chem.* **2001**, *49,* 2948-2953.
22. IDF. Sensory evaluation of dairy products by scoring-reference methodology, International IDF Standard 99C. International Dairy Federation. Brussels, Belgium. 1997.
23. Qian, M.; Reineccius, G. *J. Dairy Sci.* **2003**, *86,* 770-776.
24. Contarini, G.; Povolo, M. *J. Agric. Food Chem.* **2002**, *50,* 7350-7355.
25. Marsili, R.T. *J. Chromatographic Sci.* **1999**, *37,* 17-23.
26. Neeter, R.; de Jong, C.; Teisman, H.G.J.; Ellen, G. In *Flavor Science Recent Developments;* Taylor, A.J.; Mottram, D.S. Eds.; The Royal Society of Chemistry: Cambridge, UK, 1996; pp 293-296.
27. Fabre, M.; Aubry, V.; Guichard, E. *J. Agric. Food Chem.* **2002**, *50,* 1497-1501.
28. Qian, M.; Reineccius, G. *J. Dairy Sci.* **2002**, *85,* 1362-1369.
29. Zhentbauer, G.; Reineccius, G.A. *Flav. Fragr. J.* **2002**, *17,* 300-305.
30. Pinho, O.; Ferreira, I.M.P.L.V.O.; Ferreira, M.A. *Anal. Chem.* **2002**, *74,* 5199-5204.
31. Frank, D.C.; Owen, C.M.; Patterson, J. *Lebensm.-Wiss. u.-Technol.* **2004**, *37,* 139-154.
32. Parliament, T.H.; McGorrin, R.J. In *Flavor Chemistry: Industrial and Academic Research;* Risch, S.J.; Ho, C.T., Eds.; American Chemical Society: Washington D.C., 2000; pp 44-71.
33. McGorrin, R.J. In *Food Flavors and Chemistry: Advances of the new millennium.* Spanier, A.M.; Shahidi, F.; Parliment, T.H.; Mussinan, C.J.; Ho, C.T.; Contis, E.T., Eds.; Royal society of Chemistry: Cambridge UK, 2001; pp 67-84.
34. Moio, L.; Langlois, D.; Etievant, P.X.; Addeo, F. *Ital. J. Food Sci.* **1993**, *3,* 227-237.

35. Van Ruth, S. *Biomol. Eng.* **2001**, 17, 121-128.
36. Sen, A.; Laskawy, G.; Schieberle, P.; Grosch, W. *J. Agric. Food Chem.* **1991**, *39,* 757-759.
37. Mills, O. E.; Broome, A. J. In *Flavor Analysis: Development in Isolation and Characterization.* Mussinan, C. J.; Morello, M. J., Eds.; American Chemical Society: Washington, DC, 1998; pp 85-91.
38. Van den Dool, H.; P. Kratz. *J. Chromatogr.* **1963**, *11,* 463-471.
39. Hall, G.; Andersson, J. *J. Food Qual.* **1985**, *7,* 237-253.
40. Ulberth F.; Roubicek F. *Int. Dairy J.* **1995,** *5,* 523-531.
41. Badings, H.T.; Neeter, R. *Neth. Milk Dairy J.* **1980**, *34,* 9-30.
42. Karagül-Yüceer K.; Drake M.A.; Cadwallader K.R. *J. Sensory Stud.* **2004**, *19,* 1-14.
43. Thomsen, M.K.; Lauridsen, L.; Skibsted, L.H.; Risbo, J. *J. Agric. Food Chem.* **2005**, 53, 1805-1811.
44. Shiratsuchi, H.; Shimoda, M.; Imayoshi, K.; Noda, K.; Osajima, Y. *J. Agric. Food Chem.* **1994**, *42,* 984-988.
45. Jenness, R. In *Fundamentals of Dairy Chemistry, 3rd Ed.*; Wong, N.P., Ed.; Van Nostrand Reinhold Co.: New York, NY, 1988; pp 1-38.
46. Swaisgood H.E. In *Food Chemistry, 3rd Edition;* Fennema, O.R., Ed. Marcel Dekker: New York, NY, 1996; pp 841-876.
47. Lee, Y.B.; Morr, C.V. In *Lipids in Food Flavors.* Ho, C.; Hartman, G., Eds.; American Chemical Society: Washington, DC, 1994.
48. Creamer, L.K.; MacGibbon, A.K.H. *Int. Dairy J.* **1996**, *6,* 539-568.
49. Sienkiewicz, T.; Riedel, C. *Whey and Whey Utilization Possibilities for Utilization in Agriculture and Foodstuffs Production;* Buersche Druckerei: Leipzig, Germany, 1990.
50. Chandan, R. *Dairy-Based Ingredients;* Eagan Press Handbook Series: St. Paul, MN, 1997.
51. Jensen, R.G.; Ferris, A.M.; Lammi-Keefe, C.J. *J. Dairy Sci.* **1991**, *74,* 3228-3243.
52. Rychlik M.; Schieberle P.; Grosch W. *Compilation of thresholds, odor qualitities, and retention indices of key food odorants.* Deutsche Forschungsanstalt fur Lebensmittelchemie and Instiut fur Lebensmittelchemie der Technischen Universitat Munchen: Garching, Germany, 1998.
53. Karagul-Yüceer, Y.; Drake, M.A.; Cadwallader, K.R. *J. Food Sci.* **2003**, *68,* 1215-1219.
54. Carunchia Whetstine, M.E.; Drake, M.A. In *Handbook of Milk of Non-Bovine Mammals.* Haenleine, G.; Park, Y., Eds.; Blackwell Publishing: Ames IA, 2005.
55. Dimick, P. *Canadian Institute Food Tech. J.* **1982**, *15,* 247:256.
56. Aurand, L. W.; Boone, N. H.; Giddings, G. G. *J. Dairy Sci.* **1976**, *60,* 363-369.

57. Gilmore, T. M.; Dimick, P. S. Photochemical changes in major whey proteins of cow's milk. *J. Dairy Sci.* **1979**, *62,* 189-194.
58. Cadwallader, K.R.; Howard, C.L. In *Flavor Analysis Developments in Isolation and Characterization.* Mussinan, C.J.; Morello, M.J., Eds.; American Chemical Society: Washington, DC, 1998; pp 343-358.
59. Jenq, W.; Bassette, R.; Crang, R. E. Effects of light and copper ions on volatile aldehydes of milk and milk fractions. *J. Dairy Sci.* **1988**, *71,* 2366-2372.
60. Fox, P.F. In *Dairy Processing Improving Quality.* Smit, G., Ed.; Woodhead Publishing Limited and CRC Press LLC: Boca Raton, FL, 2003; pp 5-38.
61. Badings, Herman T. In *Volatile Compounds in Foods and Beverage.* Marse, I., Ed. Marcel Dekker Inc: New York, NY, 1991.
62. Moio, L.; Langlois, D.; Etievant, P.X.; Addeo, F. Powerful odorants in bovine, ovine, caprine, and water buffalo milk determined by means of extract dilution sniffing analysis. *J. Dairy Res.* **1993**, 60, 215-222.
63. Calvo, M.M.; de la Hoz, L. *Int. Dairy J.* **1992**, *2,* 69-81.
64. Celestino, E.L.; Iyer, M.; Roginski, H. *Int. Dairy J.* **1997**, *7,* 129-140.
65. Dimick, P.S.; Walker, N.J.; Patton, S. *J. Agric. Food Chem.* **1969**, *17,* 649-655.
66. Bendall, J.G. *J. Agric. Food Chem.* **2001**, *49,* 4825-4832.
67. Scanlan, R.A.; Lindsay, R.C.; Libbey, L.M.; Day, E.A. *J. Dairy Sci.* **1968**, *51,* 1001-1007.
68. Walstra, P.; Geurts, T.J.; Noomen, A.; Jellema, A.; van Boekel, M.A.J.S. In *Dairy Technology Principles of Milk Properties and Processes;* Marcel Dekker: New York, NY, 1999; pp 445-470.
69. Allen, C.; Parks, O.W. *J. Dairy Sci.* **1975**, *58,* 1609-1611
70. Lloyd, M.A.; Zou, J.; Ogden, L.V.; Pike, O.A. *J. Food Sci.* **2004**, *69,* 326-331.
71. Driscoll N.R.; Brennand C.P.; Hendricks D.G. *J Dairy Sci.* **1985**, *68,* 1931-1935.
72. Ferretti, A.; Flanagan, V. P. *J. Dairy Sci.* **1971**, *54,* 1764-1768.
73. Min, D.B.; Tickner, D.B.; Lee, S.H.; Reincessius, G.A. *J. Food Sci.* **1990**, *55,* 401-403.
74. Harwalker, V.R.; Cholette, H.; McKellar, R.C.; Emmons, D.B. *J. Dairy Sci.* **1993**, *76,* 2521-2527.
75. Lee, K.D.; C.G. Lo, J.J.; Warthesen. *J. Dairy Sci.* **1996**, *79,* 1521-1528.
76. Singh, T.K.; Young, N.D.; Drake, M.A.; Cadwallader, K.R. *J. Agric. Food Chem.* **2005**, *53,* 1185-1189.
77. Hammond, E. G. The flavors of dairy products. In *Flavor Chemistry of Lipid Foods;* Min, D. B., Smouse, T. H., Eds.; The American Oil Chemists' Society: Champaign, IL, 1989; pp 222-236.
78. De Block J.M.; Merchiers L.; Mortier A.; Braekman W.; Ooghe P.; Renterghem RV. *Int. Dairy J.* **2003**, *13,* 87-94.

79. BeMiller, J.N.; Whistler, R.L. In *Food Chemistry, 3rd Edition;* Fennema, O.R. Ed.; Marcel Dekker: New York, NY, 1996; pp 175-176.
80. Whitfield, F. B. *Crit. Rev. Food Sci. Nutr.* **1992**, *31,* 1-58.
81. Laye, I.; Karleskind, D.; Morr, C.V. In *Food Flavors: Generation, Analysis, and Process Influence;* Charalambous, G., Ed. Elsevier: Amsterdam, 1995; pp 769-783.
82. Creamer, L.K. *New Zealand J. Dairy Sci. Technol.* **1976**, *11,* 30-39.
83. Carunchia Whetstine, M.E.; Cadwallader, K.R.; Drake, M.A. *J. Agric. Food Chem.* **2005**, *53,* 3126-3132.
84. Carunchia Whetstine, M.E.; Croissant, A.E.; Drake, M.A. *J. Dairy Sci.* **2005**, Submitted.
85. Avsar, Y.; Karagul-Yüceer, Y.; Drake, M.A.; Singh, T.; Yoon, Y. Cadwallader, K.R. *J. Dairy Sci.* **2004**, *87,* 1999-2010.
86. Forss, D. A. *J. Dairy Res.* **1979**, *46,* 691-706.
87. Lindsay, R.C. In *Food Chemistry, 3rd Edition;* Fennema, O.R., Ed.; Marcel Dekker: New York, NY, 1996; pp 756-758.
88. Caudle, A.D.; Yoon, Y.; Drake, M.A. *J. Food Sci.* **2005**, In press.
89. Lee, Y.B.; Laye, I.; Kim, Y.D.; Morr, C.V. *Int. Dairy J.* **1996,** *6,* 485-496.
90. Ott, A.; Fay, L.B.; Cahintreau, A. *J. Agric Food Chem.* **1997**, *45,* 850-858.
91. Parks, O.W.; Schwartz, D.P.; Keeney, M. *Nature.* **1964**, *202,* 185-187.
92. Ramshaw, E.H.; Dunstone, E.A. *J. Dairy Res.* **1969**, *36,* 215-223.
93. Shibamoto, T. In *Thermal Generation of Aromas.* Parliment, T.H.; McGorrin, R.J.; Ho, C., Eds.; American Chemical Society: Washington, DC, 1989.
94. Foucaud, C.; Juillard, V. *J.Dairy Res.* **2000**, *67,* 233-240.
95. Schieberle, P. *J. Agric. Food Chem.* **1995**, *43,* 2442-2448.
96. Zhang, Y.; Ritter, W.J.; Barker, C.C.; Traci, P.A.; Ho, C.T. In *Lipids in Food Flavors;* Ho, C.T.; Hartman, T.G., Eds.; American Chemical Society: Washington D. C., 1994; pp 49-60.
97. Hidalgo, F.J.; Kinsella, J.E. *J. Agric.Food Chem.* **1989**, *37,* 860-866.
98. Mills, O.E. In *Food Flavors, ingredients, and composition: Proceedings of the 7th International Flavor Conference;* Elsevier: New York, NY, 1993.
99. Randhir, R.; Shetty, K. *Process Biochem.* **2004**, *40,* 1701-1705.
100. Suriyaphan, O.; Drake, M.A.; Chen, X.Q.; Cadwallader, K.R. *J. Agric. Food Chem.* **2001**, *49,* 1382-1387.
101. Ha, J.K.; Lindsay, R.C. *J. Dairy Sci.* **1993**, *76,* 677-690.
102. Carunchia Whetstine M.E.; Karagul-Yuceer, Y.; Avsar, Y.K.; Drake, M.A. *J. Food Sci.* **2003**, *68,* 2441-2447.
103. El Soda, M.; Law, J.; Tsakalidou, E.; Kalantzopoulos, G. In *Food Flavors, Generation Analysis and Process Influence.* Charalambous, G., Ed.; Elsevier Science: New York, NY, 1995; pp 1823-1847.
104. Milo, C.; Reineccius, G.A. *J. Agric. Food Chem.* **1997**, *45,* 3590-3594.

105. Cobb, W.Y.; Patton, S.; Grill, H. *J. Dairy Sci.* **1963,** *46,* 566-567.
106. Shipe, W. F.; Bassette, R.; Deane, D. D.; Dunkley, W. L.; Hammond, E. G.; Harper, W. J.; Kleyn, D. H.; Morgan, M. E.; Nelson, J. H.; Scanlan, R. A. *J. Dairy Sci.* **1978**, *61,* 855-869.
107. Tomaino, R. T.; Parker, J. D.; Larick, D. K. *J. Agric.Food Chem.* **2001**, *49,* 3993-3998.

Chapter 14

Effect of Processing Technology and Phenolic Chemistry on Ultra-High Temperature Bovine Milk Flavor Quality

Devin G. Peterson, Paula M. Colahan-Sederstrom, and Rajesh V. Potineni

Department of Food Science, The Pennsylvania State University, 215 Borland Laboratory, University Park, PA 16802

Dairy ingredients/products utilize a thermal processing step in the production cycle as a means of improving food preservation, product safety and ease of distribution. However, for specific dairy-based materials, thermal treatment can also result in the development of negative product traits due to the simultaneous generation of aroma compounds. For example, ultra-high temperature (UHT) processed milk, although commercially sterile, has a 'cooked' or 'stale' aroma that is often considered a significant product defect. Conversely, thermal processing can also have positive affects on flavored dairy-based due to enhanced flavor stability. Various modes of flavor control in UHT fluid milk are discussed.

Flavor is a key attribute of any food product, as it is one of the major factors influencing a consumer's choice of food (*1*). The desired flavor of fluid milk is a mild, bland, or low flavor intensity product. Even a very slight change in the flavor profile of milk can be unacceptable to the consumer (*2, 3*). According to Lampert (*4*), milk consumers who drink raw milk can tell a difference in the flavor properties of high temperature short time (HTST) pasteurized milk. In the United States, milk consumers have become accustomed to pasteurized milk (mandated by law) and consequently the flavor of pasteurized milk is both accepted and expected. Consequently, the cooked off-flavor notes in ultra-high temperature (UHT) processed milk has not been accepted by American consumers even though the product has the distinct advantage of shelf-stability. More recently, new thermal processing technologies are being used to minimize off-flavor problems of traditional ultra-high temperature processing such as extended shelf-life (ESL) milk. However, the flavor properties of both UHT and ESL milks are considered to be lower in quality (liking) in comparison to pasteurized milk, which makes UHT or ESL milk less acceptable to the consumer (see Figure 1). Research preformed at Cornell University and sponsored by New York State Milk Promotion Order, a division of the New York Department of Agriculture and Marketing, showed a direct correlation between flavor quality of milk and level of consumption (*5*). The negative flavor properties of shelf-stable fluid milk can be primarily associated with chemical changes induced from the Maillard reaction, protein degradation, and lipid oxidation/degradation reactions.

There are two main commercial thermal processing methods for the production of shelf-stable milk products, both direct and indirect techniques, to obtain commercial sterility. Both the direct and indirect methods produce similar milk products in stability, microbial safety and shelf-life, however, there is a distinct difference between these thermal treatments on the product flavor attributes. In the direct heating systems, super-heated steam is either sprayed or injected into the raw milk which heats the milk very quickly. During this process the milk volume is increased by approximately 11%, but the water is vaporized and removed during a rapid cooling stage via a vacuum chamber (*7-9*). For the more conventional indirect heating system, milk is heated with a heat exchanger (tubular or plate) and this system is considered to be more economical than direct systems (less expensive for both initial costs and running costs). The indirect system can have regeneration levels of up to 90% of thermal energy whereas for the direct processing system is less effective in energy regeneration (~50%). Thus the total cost for the direct processing system can be twice as high as an indirect system (*9*). Nonetheless, the heating rate of the indirect of system is much slower than the direct heating system and as a result the major benefit of the direct system comes from the less severe heat treatment (overall lower thermal dose), and therefore less flavor development (*8*).

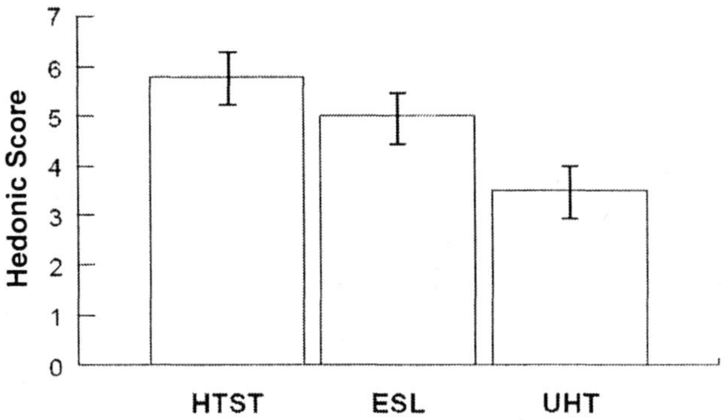

Figure 1. Hedonic Sensory Score Versus Extent of Heat Treatment of Milk; Average consumer acceptance (hedonic rating with 1 = extremely dislike, 9 = extreme like) comparing HTST (74°C for 4 seconds), ESL (134°C for 4 seconds, direct steam injection) and UHT (indirect heating, plate-exchanger) processed milk (Adapted with permission from reference 6. Copyright 1995 International Association of Food Protection)

In addition to the application of processing technology to improve the flavor properties of shelf-stable milk, the use of natural product chemistry has been recently applied to reduce the off-flavor attributes of UHT milk. Based on recent findings of Peterson and Totlani (*10*) who reported that specific flavonoids can alter (reduce) Maillard-type flavor generation pathways, Colahan-Sederstrom and Peterson (*11*) investigated the ability of epicatechin (a flavonoid) to reduce the thermal development of aroma compounds (i.e., Maillard reaction products) formed during ultra-high temperature (UHT) processing (indirect system) of bovine milk. Colahan-Sederstrom and Peterson reported that epicatechin (EC) added to raw fluid milk prior to UHT processing reduced the thermal generation of aroma compounds (cooked flavor). A direct comparison of the aroma properties between the a traditional UHT milk product and UHT milk with 0.1% EC added prior to thermal processing is reported in Figure 2 (based on AEDA; any odorants which reported an FD difference ≥ 4 between the control and treatment milk samples are illustrated). Based on the direct comparison of the FD-factors between the control and treatment milk samples, the largest impact the addition of EC had on off-flavor development during UHT processing was reported for methional (a potent cooked-type note). The concentration of methional in the treatment sample was estimated to be approximately 32-fold lower in the control sample, based on the observed FD-value ratio between the control and treatment sample of 32. The compound, 2-isoproypl-3-methoxypyrazine was tentatively identified and is not a typical Maillard reaction product (known microbial conversion product) but was listed

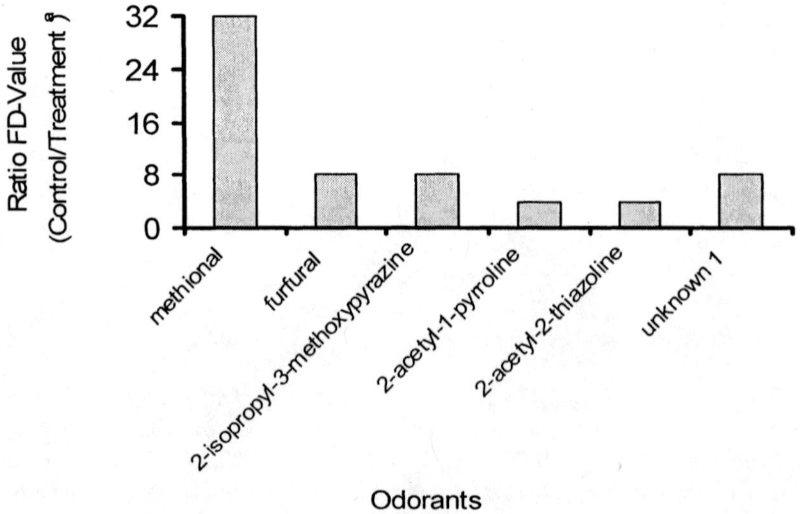

Figure 2. Ratio of flavor dilution (FD) values of control versus treatment for each odorant with a difference ≥ 4; a = $(2^{FD\ control})/(2^{FD\ treatment})$; (Adapted with permission from reference 11. Copyright 2005. American Chemical Society)

as it has been previously reported in milk powder (*12, 13*). The sensory analysis of the cooked flavor intensity for these UHT samples (see Table 1) was in agreement with the AEDA data illustrated in Figure 2 as the treatment sample was found to be statistically lower in cooked flavor intensity in comparison to the control sample. Polyphenolic compounds are commonly associated with bitterness and likewise food rejection, although at the 0.1% level, bitterness intensity was determined not to be statically different form a conventional UHT milk sample (Table I).

Table I: Mean Scores for Cooked Flavor and Bitterness in Milk Samples[a]

Milk Sample	*Cooked Flavor* [1,3] *(LSD = 1.61)*	*Bitterness* [2,3] *(LSD = 1.78)*
Control UHT	5.11 A	0.19 B
Treatment 0.1% EC UHT	2.88 B	0.70 B

[1] = 15 centimeter scale was used with 0 = no detectable cooked (pasteurized) and 15 = very cooked (n=10). [2] A 15 centimeter scale was used with 0 = no bitterness and 15 = very bitter (n=9). [3] Means in the same column having the same letter are not significantly different (α = 0.05); (Adapted with permission from reference 11. Copyright 2005. American Chemical Society)

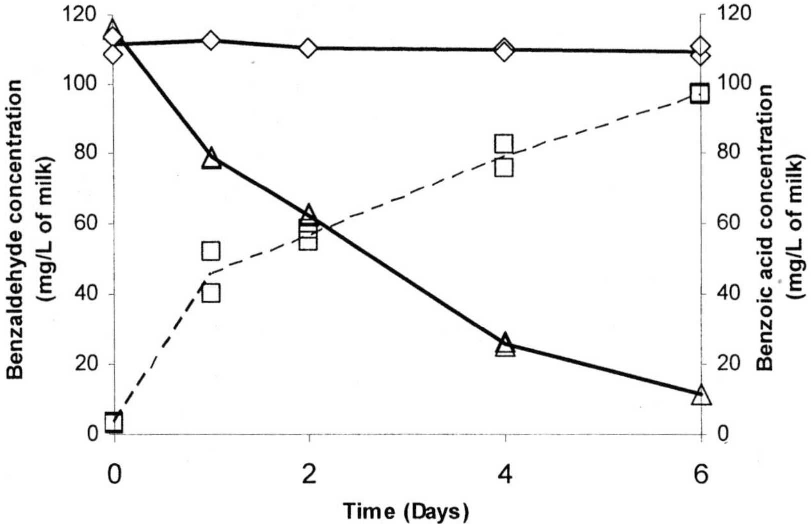

Figure 3. Effect of whole milk thermal treatment on benzaldehyde stability storaged at 5°C under aerobic conditions; △ = benzaldehyde (pasteurized milk sample), ◇ = benzaldehyde (UHT milk sample), □ = benzoic acid (pasteurized milk); (Adapted with permission from reference 23. Copyright 2005. American Dairy Science Association)

Although thermal processing can impart negative product attributes (i.e. cooked flavor), heat can also positively affect the flavor stabilitity of flavor enriched dairy-based food products (ice cream, flavored milks, yogurt, etc). Flavor degradation reactions in dairy-based products has been linked to enzymatic reactions (*14-21*) and consequently flavor stability has been positively correlated to the extent of the thermal dose during processing (or level of enzymatic inactivation). Both Anklam et al. (*16*) and Gassenmeier (*17, 22*) studied the degradation of vanillin to vanillic acid in select dairy products during storage and associated this degradation reaction to oxidative activity of the intrinsic milk enzyme xanthine oxidase. In a recent study, Potineni and Peterson (*23*) investigated the influence of milk thermal processing conditions (or potential enzyme inactivation) on benzaldehyde stability in fluid milk. These authors reported that when benzaldehyde was spiked in pasteurized milk, over 90% of this compound was oxidatively converted to benzoic acid after a 6-day storage period, while in UHT mik, benzaldehyde was found to be completely stable (no loss observed as well as no formation benzoic acid) over an equivalent storage time (see Figure 3). Xanthine oxidase (XO), which has been previously suggested to degrade vanillin would be inactivated in UHT processed milk (*24*).

However, when these authors spiked XO in UHT milk at typical intrinsic levels found of raw milk, benzaldehyde degradation was not reported. Furthermore, no autooxidation products (i.e. hexanal) were reported in the stored samples. Based on these findings, the degradation of benzaldehyde in milk is likely enzymatic (a dehydrogenase) and probably requires NADH or NADPH which may be more effectively bound by milk proteins in higher heat treated milk products (higher degree of denaturation).

In summary, thermal processing can have both negative and positive implications on the flavor properties of dairy-based foods or beverages (i.e. UHT milk) and ultimately the product quality. Both processing technology and the application of phenolic chemistry have shown potential to reduce unwanted cooked notes in shelf-stable milk and as such may provide dairy processors the ability to produce low cooked dairy products with improved stability for flavored dairy products.

References

1. Glanz, K.; Basil, M.; Maibach, E.; Goldberg, J.; Snyder, D. *J. Am. Diet. Assoc.* **1998**, *98*, 1118-1126.
2. Gonzalez-Cordova, A.F.; B. Vallejo-Cordoba. *J. Agric. Food Chem.* **2003**, *51*, 7127-7131.
3. Dumont, J.P.; Adda, J. in *Progress in Flavour Research*, D.G. Land, Ed.; 1979, Applied Science Publishers LTD: London. p. 245 - 262.
4. *Modern Dairy Products*. Lampert, L.M., Ed.; New York, Chemical Publishing Company, Inc. 1965.
5. Bandler, D.K. *Dairy Food Sanit.* **1982**, 2:312-315.
6. Blake, M.R.;Weimer, B.C.; Mcmahon, D.J.; Savello, P.A. *J. Food Protect.* **1995**, *58*, 1007-1013.
7. *Continuous Thermal Processing of Foods: Pasteurization and UHT Sterilisation.* Lewis, M.; Heppell, N. Eds.; 2000, Gaithersburg, Maryland: Aspen Publishers, Inc.
8. Datta, N.; Elliott, A. J.; Perkins, M. L.; Deeth, H. C. *Aust J. Dairy Technol.* **2002**, *57*, 211-227.
9. *Ultra-High-Temperature Processing of Milk and Milk Products.* Burton, H. Ed.; London, New York: Elsevier Applied Science. 1988.
10. Peterson, D.G.; Totlani, V.M. in *Phenolics in Foods and Natural Health Products*, F. Shahidi; C.-T. Ho, Eds. 2005, ACS: Washington, DC. pp 143-160.
11. Colahan-Sederstrom, P.M.; Peterson, D.G. *J. Agric. Food Chem.* **2005**, *53*, 398-402.
12. Shiratsuchi, H.; Shimoda, M.; Imayoshi, K.; Noda, K.; Osajima, Y. *J. Agric. Food Chem.* **1994**, *42*, 984-988.

13. Karagul-Yuceer, Y.; Drake, M. and Cadwallader, K. *J. Agric. Food Chem.* **2001**, *49*, 2948-2953.
14. Jeon, I.J. in *Food taints and off-flavors*, M.J. Saxby, Ed. Chapman & Hall: London, UK. 1993, pp 139-167.
15. Chevalier, M.; Prat, Y.;Navellier, P. *Ann. Fals. Exp. Chim.* **1972**, *697*, 12-16.
16. Anklam, E.; Gaglione, S.; Muller, A. *Food Chem.* **1997**, *60*, 43-51.
17. Gassenmeier, K. Lebensm. *Wiss. U,-Technol.* **2003**, *36*, 99-103.
18. Baumgartner, J.; Neukom, H. *Chimia.* **1972**, *26*, 366-368.
19. Kempe, K.; Kohnen, M. *Adv. Food Sci.* (CMTL), **1999**, *21*, 48-53.
20. Allen, J.C.; Wrieden, W.L. *J. Dairy Res.* **1982**, *49*, 249-263.
21. Ostdal, H.; Bjerrum, M.J.; Pedersen, J.A.; Andersen H.J. *J. Agric. Food Chem.* **2000**, *48*, 3939-3944.
22. Gassenmeier, K., in *Handbook of Flavor Characterization: sensory analysis, chemistry, and physiology*, K.D. Deibler and J. Delwiche, Eds.; Marcel Dekker, Inc.: New York. 2004, pp 259-266.
23. Potineni, R.V.; Peterson, D.G. *J. Dairy Sci.* **2005**, *88*, 1-6.
24. *Dairy technology: principles of milk properties & processes.* Walstra, P.; Geurts, T.J.; Noomen, A.; Jellema, A.; Van Boekel, M.A.J.S. Eds.; New York, NY: Marcel Dekker Inc. 1999, pp 91-99.

Indexes

Author Index

Subject Index

B

C

D

E

F

G

M

N

O

R

S

T